ADVANCE PRAISE

Green Building A to Z

The growth of Environmental Design is staggering. Jerry Yudelson continues his enormous contribution to this movement with *Green Building A to Z.* This book is a must read for any emerging green building enthusiast.

— JOE VAN BELLEGHEM,
Managing Partner Windmill Development Group Ltd.,
Founder Canada Green Building Council

In this very useful reference guide, Jerry mixes straightforward information with his own personal and engaging perspective that makes this document beneficial for all those interested in green building from novice to experienced.

— MARY ANN LAZARUS,
AIA, LEED® AP, Director of Sustainable Design,
HOK Architects

An outstanding resource that collects very diverse and timely information into a "Green Encyclopedia" that should be on everyone's bookshelf, *Green Building A to Z* is an easy-to-read guide that brings it all together under one cover.

— ROD WILLE, Sr.
Vice President, Manager of Sustainable Construction,
Turner Construction Company

Yudelson brings together the basics of green building and adds to it something for everyone, even the most experienced green champion. *Green Building A to Z* does not belong on your library shelf. It's an everyday tool!

— DR. KATH WILLIAMS,
Past President, World Green Building Council

Green Building A to Z provides a compelling reference guide that articulates the key terms and concepts in green building. In this fast-changing field, Jerry Yudelson's lucid writing style beautifully describes the essential concepts and terminology with accuracy and passion. *Green Building A to Z* creates a much-needed map of the green building landscape.

— ANDRÉS R. EDWARDS,
author of *The Sustainability Revolution: Portait of a Paradigm Shift*,
and founding board member, US Green Building Council's
Redwood Empire Chapter

Green Building A to Z is a wonderful resource for professionals learning about the green building movement and its many moving parts. Yudelson does an excellent job pulling so many different topics into one book. I recommend this book to anyone starting their journey on the path to sustainability.

— PAUL SHAHRIARI,
founder, GreenMind

Green Building A to Z is an exciting contribution to the tools we need to create environmentally intelligent communities. As the sustainable building industry explodes, it is vital that we have a common language and shared expectations about what constitutes "green." Jerry Yudelson has produced a dictionary of terms that is understandable to the layperson and professional alike. It will be especially pertinent to those just beginning to understand the dynamic new paradigm of sustainable design.

— RON VAN DER VEEN,
AIA, LEED® AP, Principal,
MITHUN architects + designers + planners

Green Building A to Z provides thorough coverage of the language and practice involved in the design of green buildings. Yudelson gives us compelling and illustrated definitions used in sustainable design, along with his expertise, historical and philosophical context.

— ALISON G. KWOK,
AIA, LEED® AP, Associate Professor, Department of Architecture,
University of Oregon

Green Building A to Z

Understanding the Language of Green Building

Jerry Yudelson

foreword by **Kevin Hydes**

New Society Publishers

Cataloging in Publication Data:
A catalog record for this publication is available from
the National Library of Canada.

Cover design by Diane McIntosh. Images: Alamy and iStock.

Printed in Canada.
Second printing January 2008.

Paperback ISBN: 978-0-86571-572-1

To order directly from the publishers, please call toll-free
(North America) 1-800-567-6772, or order online at www.newsociety.com

Any other inquiries can be directed by mail to:

New Society Publishers
P.O. Box 189, Gabriola Island, BC V0R 1X0, Canada
(250) 247-9737

New Society Publishers' mission is to publish books that contribute in fundamental ways to building an ecologically sustainable and just society, and to do so with the least possible impact on the environment, in a manner that models this vision. We are committed to doing this not just through education, but through action. We are acting on our commitment to the world's remaining ancient forests by phasing out our paper supply from ancient forests worldwide. This book is one step toward ending global deforestation and climate change. It is printed on acid-free paper that is 100% post-consumer recycled (100% ancient forest-free), processed chlorine free, and printed with vegetable-based, low-VOC inks. Additionally, New Society purchases carbon offsets based on an annual emissions audit, operating with a carbon-neutral footprint. For further information, or to browse our full list of books and purchase securely, visit our website at: www.newsociety.com

NEW SOCIETY PUBLISHERS www.newsociety.com

Contents

Part III

For green builders
and their supporters,
everywhere.

Preface

This book presents the basic concepts and terminology used in designing and constructing green buildings, based on state-of-the-art design and construction practices in 2007. It is designed for you, the intelligent reader, who may not be actively engaged in architecture or building engineering, but who needs a quick introduction to the rationale for green buildings and the language of the field. It will also be useful for public officials; for those dealing with green building or sustainability requirements from within or outside your company, organization or agency; for those whose livelihood depends on financing, building and marketing commercial development projects, and new residential subdivisions or multifamily projects; for real estate brokers and agents; for people in the finance, insurance and real estate industries; for senior executives in universities, government agencies and large corporations who want to understand what all the fuss is about; and for anyone who has an interest in turning the design, construction and operation of buildings into a more environmentally responsible activity.

Green building enthusiasts have made this phenomenon a major part of the design and construction industry over the past ten years. Since 2000 the number of green buildings has grown from a handful to more than 5,000 projects in the US and over 400 in Canada actively seeking certification of one kind or another at the end of 2006.[1] This is the fastest-growing phenomenon to hit the building industry since the Internet and perhaps since air conditioning. I saw a need for a quick and accessible guide to green buildings that could be used by a wide variety of people, one that is technically accurate and up-to-date, without requiring a professional or technical background.

The following chapters address several key questions: What is a green building? Why are green buildings and green developments important for the environmental and economic challenges we face early in the 21st century? What are the important new sustainable technologies that are influencing the building industry? How do green products and green buildings actually get designed and built? What can I do in my company, my home and my city to further the "green building revolution"?

Throughout the text, I rely on published data, current through early 2007. Most of this information is available from sources such as the US and Canada Green Building Councils, from papers at green building conferences, from the Internet or from business or trade media. I have examined several public and proprietary surveys, and I have benefited from personal conversations with green building leaders to round out the roster of topics important to understanding green buildings, green products and green developments in the United States.

I have spent most of my career engaged in energy and environmental affairs, working to make our current economy and way of life more appropriate to long-term sustainability. As a student in California in 1970, I helped organize the first Earth Day, and I helped create the first state-level agency to promote solar energy, also in California. For the past ten years, I have been involved in building design and construction on a daily basis, and I've been active in the green building movement since 1999. I see my role as a communicator between green building professionals and the larger business and governmental public. I conceived of this book as a way to accelerate the understanding of the importance of green buildings in addressing the climate-change challenges of the early 21st century.

Each of us has an important role to play in transforming the building and development industry into one that produces what most people say they want from it: energy- and resource-efficient, environmentally sound, healthy, comfortable and productive places to live, work, learn, experiment and play. I hope this book will help you to play a role in this great undertaking. Thanks for your interest, and may you read this book profitably! I welcome any other feedback, directed to me at: jerry@greenbuildconsult.com, or via my business website, www.greenbuildconsult.com.

I want to thank all the people who contributed to the case studies in this book and to all those toiling in the green building vineyard. Especially I want to thank my research associate, Gretel Hakanson, who came to this project late and provided invaluable assistance in pulling together the case studies, photos and examples in this book. Thanks to Lynn Parker of Parker Designs, Beaverton, Oregon, for the graphic images created especially for this book. And a very special thanks to Kevin Hydes, former chair of the US Green Building Council and chair-elect of the World Green Building Council, for generously agreeing to write the foreword. I also want to thank all the green building professionals who furnished photos and project descriptions, and especially to those experienced architects, engineers, builders and consultants who reviewed the manuscript, including Sonja Persram, Mark Heizer, Mary Ann Lazarus, Josh Arnold, Jessica Yudelson, Clark Brockman and Yves Khawam.

Jerry Yudelson, PE, MS, MBA, LEED® AP
Tucson, Arizona
March 2007

Foreword

By Kevin Hydes

At the beginning of the millennium, which now seems like a lifetime ago in terms of green building chronology, I happened to meet another engineer as we were beginning to embark on a new wave of buildings and next-generation green design. At the time I had just become president of my former firm, the Canadian-based Keen Engineering, and I was articulating the vision we had set, to change our focus from "blue" to "green," from traditional building service or mechanical engineering to a more enlightened direction.

We talked about the imperative of the design and construction community making the shift to a new paradigm and how could we do this quickly, effectively and economically. My biggest challenge was not only to train a new generation of thinkers but to find many more, as the market demand for green building know-how was beginning to explode, along with the rapid growth of our own business.

As I described my frustration at not being able to "find" enough good folks but noted that I had no difficulty in finding clients, the engineer across the table looked at me calmly and said, "Kevin, your solution is simple — recruiting and marketing are the same thing, two sides of the same coin." In an instant I realized that not only was he right, but all I needed to do was apply the same ideas and conviction in dealing with potential recruits that I was using with my clients. It worked: our firm tripled in size in five years and became much more profitable.

That engineer's name was Jerry Yudelson. Jerry has a unique gift, one that few of us possess, to take a series of complex and often conflicting data, make sense of it, then boil the message down to its essence. In my opinion, he is one of the great communicators of our time.

In recent months, thanks to a confluence of events, we have seen the momentum build globally around a shared concern for the future of our planet. Climate change has shifted from being a purely scientific discussion to a mainstream concern in a short period of time. Even in recent weeks, we have seen new information from the Intergovernmental Panel on Climate Change confirming the part that humanity has had in creating this problem. Business and government leaders returned from the 2007 World Economic Forum in Davos, Switzerland, united in their resolve to lead the fight against climate change. This is an historic moment.

We now know that residential and commercial buildings are the biggest single contributor to producing carbon dioxide emissions, intimately linked with global warming. At the heart of the building industry are designers, builders, developers and product manufacturers who are now committed to working together to change the way we do business. As former chair of the US Green Building Council and now incoming chair of the World Green Building Council, I have had an opportunity to observe how industry and government are coming together to dramatically reduce the impact of buildings on the environment, using new technologies and systems that help in reducing carbon dioxide emissions, improving the quality of stormwater and reducing the habitat destruction caused by urban growth.

This book is a valuable resource for those who want to know more about the full range of issues tackled by the green building movement. It weaves the global issues, the historic perspectives and precedents for green buildings, current and emerging technology and trend data that lead the reader, not only in understanding the principles and business case for shifting to green practice, but also in shifting the mindset from service provider or supply-chain player, to concerned and knowledgeable advocate.

I often ask people, "Who was the greatest engineer — Thomas Edison or Henry Ford?" For me, the answer is "both": Edison, the greatest inventor of his time, and Ford, the great replicator, the industrialist. In the late 19th century, Edison developed many inventions that led us into a new era of technological advancement. He created the first industrial research laboratory that systematically looked for solutions to pressing problems. Ford took some of Edison's inventions, as well as those of Harvey Firestone (tires) and others, and focused on replication, refinement and simplification, so that we could all afford the inventions through mass production. Nearly 100 years ago, Ford developed the modern system of mass production that benefits all of us to this day.

Today we need to take the innovations created by many architects and engineers on a building-by-building basis in every region of the country and around the world, then replicate these best practices rapidly throughout the built environment. Written in simple language, easily accessible to the non-specialist, and backed up by data and common sense, this book is a platform for aiding that replication, allowing us to shift to greening our cities and communities from just designing one building at a time, making first one organization at a time respond to the need for sustainable design and development, and finally leading to one "green city" at a time, until we have completely accomplished this green building revolution.

This book is all about ideas, proven and undeniable. From my own experience, I know that to effect massive change we need to take ideas and act upon them to be successful. The call for action is now. Thank you for this gift, Jerry.

Kevin Hydes, PE, P.Eng.
Montreal, Quebec
February 2007

Kevin Hydes is Vice President, Stantec Consulting, Ltd., Canada; Founder, Canada Green Building Council; and Chairman, World Green Building Council.

PART I

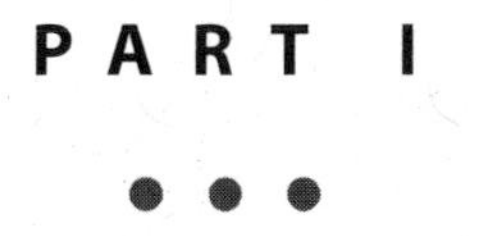

Green Buildings Today

CHAPTER 1

Green Buildings in a Global Context

Green buildings are part of a global response to increasing awareness of the role of human activity in causing global climate change. Buildings account for more than 40% of all global carbon dioxide emissions, one of the main culprits implicated in the phenomenon of global warming. While the US, Western Europe, Canada and Japan contribute the majority of greenhouse gas emissions at the present time, this situation is going to change dramatically in the near future. The projected rapid growth of carbon dioxide emissions from China, India, the rest of Asia, Brazil and Russia make it imperative that the entire world participate in reducing the "carbon footprint" (the impact on the environment in terms of the amount of greenhouse gases produced, measured in units of carbon dioxide) of urban civilization over the next 30 years. Global temperature increases are inevitable, with significant consequences for all of us.

Many observers predict that half the new building over the next three decades will occur in China alone, some 220 billion square feet of new space for residential, commercial and industrial uses. Without a focus on energy-efficient and green buildings, we have no chance for tackling global climate change. The US and other developed countries need to set a leadership example in tackling their own contributions to carbon dioxide emissions. There is every reason to believe that this is not only the socially responsible thing to do, but that it will be good business as well, for the entire world soon will be buying all manner of "carbon reduction" technologies. As the most inventive nation in the world, the US is poised to reap enormous economic advantage from addressing climate change issues in buildings with new technologies, processes and systems. By doing so, we will not only grow our own economy but will also make a major contribution to the global problem.

There are other practical reasons for innovating with green buildings. Consider for a moment the major scarcities of clean potable water around the world, as well as worsening long-term drought conditions in the American Southwest and in places such as Australia. New water conservation, wastewater treatment and water reuse technologies promise to cut building water use in half, perhaps leading to a 5% to 10% reduction in total water use. Learning how to cut energy use in buildings will also cut water use from thermal power plants (coal and nuclear), estimated to use half the water supply in the western US, directly or indirectly.

In many parts of the world, conflicts over energy and water resources are becoming common. Global warming threatens the water supplies of much of the world dependent on summer runoff from glaciers and high-mountain snowpacks for summer irrigation. Some predict that as the Himalayas become more rainy and less snowy, and as water from their snowpack runs off faster in the spring, the entirety of Southeast Asia dependent on the Mekong River, for example, will begin to experience worsening drought conditions, along with the need to make vast infrastructure investments in water desalination, water storage and water conservation systems. Saving water in urban uses such as buildings is critical for many urbanizing areas. Water may very well assume the economic and strategic importance in the coming several decades that oil has had for nearly 100 years.

Energy shortages are already a way of life in much of the world, but more acutely so in the rapidly developing mega-cities around the globe. In fact, most of the 40 largest cities in the world are not in the currently developed world; they are places such as Rio de Janeiro and Sao Paulo, Brazil; Mumbai, Chennai, Pune, Bangalore, Delhi and Kolkata, India; Karachi and Lahore, Pakistan; Hong Kong, Beijing, Chonqing, Wuhan, Tianjin, Shenyang, Guangzhou and Shanghai, China. Of the top 40 cities by population, there are only two in the US: New York and Los Angeles. In Japan, only one: Tokyo; in Russia, only two: St. Petersburg and Moscow; in Western Europe, only London; in developed East Asia, only Seoul and Singapore.[1] Clearly, we must introduce green buildings on a massive worldwide scale to halt the growth of carbon dioxide emissions and avoid the potential for major climate changes and severe economic and health disruptions over the next 30 to 50 years.

Green buildings also present a way to attack the inequity of global resource distribution by providing affordable housing for the poor that is healthier, more resource efficient and cheaper to own and operate. Already many architects, engineers and planners have responded to the disaster of Hurricane Katrina in 2005 by developing innovative housing designs that allow poor and lower middle-class people to have a healthy, attractive home, with lower utility costs and more flood-proofing than conventional housing. Renewable energy systems using the ubiquitous solar and wind energy of the planet are powering many poor villages in the developing world, helping to provide education and healthcare in resource-poor environments.

Finally, green buildings are good for the environment. Features such as green roofs emphasize sensitivity to urban habitat preservation. Innovative onsite stormwater management and the use of sustainably harvested wood and recycled-content materials help reduce the environmental and infrastructure effects of our current building methods. The essence of good design is having one action carry multiple benefits.

In 2005 an innovative competition to design conventional housing with zero or positive resource and environmental impacts drew more than 600 entries from around the world. The winning team, a group of young designers from Mithun architects and planners in Seattle, Washington, designed a house that operates totally on renewable energy, but with a few twists, as the design team explains their concept.

> (The core of the house) extends vertically, clad with a super-conductive photosynthetic plasma-cell skin that is able to generate 200% more electrical voltage per area than contemporary photovoltaics. Building on current research involving extracted spinach protein, this living skin is photosynthetic and phototropic; it grows and follows the path of the sun, generating electricity in excess of single family needs. Excess power is distributed to neighboring homes and street lighting infrastructure.

This is an example of the type of out-of-the-box thinking that green buildings are eliciting. The design also addresses water reuse, materials selection, ventilation needs and community connectedness.[2]

New green building materials are showing how we can reduce the impact on people and ecosystems from chemicals that contain persistent bioaccumulative and toxic compounds. By applying the "precautionary principle" — in essence, putting the burden of proof on industry to test everything and know its effects fully before releasing new chemicals into the human and natural ecosystems — green building product selections can help reduce the "chemical soup" that causes acute chemical sensitivities in many people. At the larger ecosystem level, the precautionary principle is an application of the Hippocratic Oath that doctors abide by: "First, do no harm."[3] There are strong reasons to suspect that human ingenuity is not infallible and that natural systems that have evolved over millions of years without having to deal with industrial chemicals are far more fragile than we assume.

Some 40 years ago, the *Whole Earth Catalog*, a bible of sorts to many Baby Boomers, coined the slogan "We are as gods and better get good at it." The essence of that message is that if human beings are remaking the Earth in their own image — a process well underway — we'd better start drawing lessons from nature about achieving long-term sustainability on a very finite planet. Green buildings are a major priority for achieving sustainable development without sacrificing quality of life for all Earth's inhabitants, human and otherwise. The *Hannover Principles*, first enunciated in 1992 by American architect William McDonough and German chemist Dr. Michael Braungart, give clear guidance for sustainable design. The first principle reads simply: "Insist on the rights of humanity and nature to co-exist in a healthy, supportive, diverse and sustainable condition."[4] Green buildings are an organized approach to conforming to this attitude.

Jeffrey Totaro Photography

A LEED® Gold-certified Building at Penn State University, designed by Overland Partners with WTW Architects, reduces energy use by 40%.

Resource Depletion and Carbon Dioxide Emissions

According to the US Green Building Council, the annual direct impacts of all US residential and commercial buildings include 39% of total energy use, 68% of electricity consumption and 30% of greenhouse gas emissions. Add in the embodied energy in making building materials, getting them to the job site, installing and servicing them, and total energy use is closer to 48%. Buildings make a major impact on just about every aspect of the world we live in; building design and construction can account for up to 30% of raw materials use, 40% of non-industrial landfill waste (including 31% of the mercury in municipal waste); 12% of potable water use, according to the US Green Building Council and the US Environmental Protection Agency.[5] Taking firm actions to reduce the environmental impacts of buildings can have a number of beneficial effects:

- Reduce ocean and river pollution from stormwater runoff.
- Extend the life of municipal infrastructure by using less water and contributing less stormwater, thereby allowing growth without infrastructure expansion.
- Extend the life of landfills by reducing the disposal of construction debris and building materials.

Most of the buildings in this country in the year 2035 (less than 30 years from now) have yet to be built or renovated, so now's the time to make changes. Between tearing down many older buildings, renovating some that are structurally sound or architecturally significant and building new structures, most of our building stock can be influenced by actions we take today to green the built environment. The green building movement will serve to make our stock of buildings more energy- and water-efficient and less burdensome on the municipal infrastructure that we all pay for, one way or another. According to one commentator, architect Edward Mazria:

> In the year 2035, three-quarters of the built environment in the US will be either new or renovated [representing more than 300 billion square feet of construction]. This transformation over the next 30 years represents a historic opportunity for the architecture and building community to reverse the most significant crisis of modern time, climate change.[6]

CHAPTER 2

Green Building History

In the late 1980s, the American Institute of Architects (AIA) created the Committee on the Environment (COTE), which has outlets today in just about every AIA chapter across the country. All across the US and Canada, architects have led the charge toward sustainable design, working through local COTE chapters, as well as the US Green Building Council chapters.

Created in 1993, the US Green Building Council (USGBC) aims to transform the building industry into a more environmentally responsible activity. Beginning in the mid-1990s, the USGBC undertook, with financial assistance from the US Department of Energy, the development of a rating and evaluation system to define what a green building represented. The first system, dubbed Leadership in Energy and Environmental Design or LEED, for new construction and major renovations, was piloted or beta-tested in 1998 and 1999 on about 50 projects in the US. In March 2000, version 2.0 of LEED was introduced as an updated, revised and expanded version of the original LEED version 1.0. Since then version 2.0 has had two major changes; LEED for New Construction (LEED-NC) version 2.2, effective since late 2005, is the current standard.

The USGBC enjoyed rapid growth from 1998, when it had only about 100 members, to the beginning of 2007, when membership stood at more than 7,700 corporate, institutional, governmental and nonprofit organizations (it does not have individual members).[1] Representing all segments of the building industry and environmental community, the USGBC has been able to craft a consensus standard for evaluating the environmental attributes of buildings and developments, by drawing on the resources of this large ($1 trillion annual construction value) and diverse industry.

Established in 2004, the Canada Green Building Council (CaGBC) now has more than 1,300 member organizations, with chapters in many provinces.[2] The CaGBC uses the LEED evaluation system but has adapted it for Canadian conditions. By 2007 the CaGBC had more than 225 projects registered for certification under the Canadian LEED standard. Green building in Canada is a fast-growing movement, with a special focus on energy efficiency and indoor air quality suitable for a more northerly and colder climate.

Current Situation

Owners and developers of residential, commercial and institutional properties across North America are discovering that it is often possible to build green buildings on conventional budgets. Many developers, building owners and facility managers are advancing the state of the art in commercial and large residential buildings through new modeling tools, design techniques and creative use of financial and regulatory incentives. For the past ten years, in ever-increasing numbers, we have begun to see development of commercial structures using green building techniques and technologies. With more than 1,200 corporations issuing sustainability reports of some form in 2006, it is clear that this market will not be a short-lived fad. Companies want to locate in a space that reflects their values, and a high-performance building goes a long way toward satisfying that requirement.[3]

Most long-time participants in the real estate, architectural design and building construction industries realize that sustainable design is the biggest *sea change* in their business careers. The urgency of global warming and the increasing US dependence on imported fuels have led architects to urge more concerted action to reduce energy use in buildings. In late 2005 the American Institute of Architects, representing more than 70,000 architects, released a major policy statement that sets a goal of reducing the fossil fuel consumption of new buildings by 50% by the year 2010, with additional 10% reductions every five years thereafter, to reach 90% reduction from 2005 levels by 2030. While this declaration has no legal force, it does add pressure to incorporate superior energy performance into the goals for each project.[4] As architect Edward Mazria observes, one can achieve a 50% reduction with existing building technology at no extra cost by simply using the right design strategies, such as proper orientation and form, daylighting, solar control and passive heating and cooling techniques.

Understanding Green Buildings

What do we mean when we speak of green buildings or high-performance buildings? According to the USGBC, these buildings incorporate design and construction practices that significantly reduce or eliminate the nega-

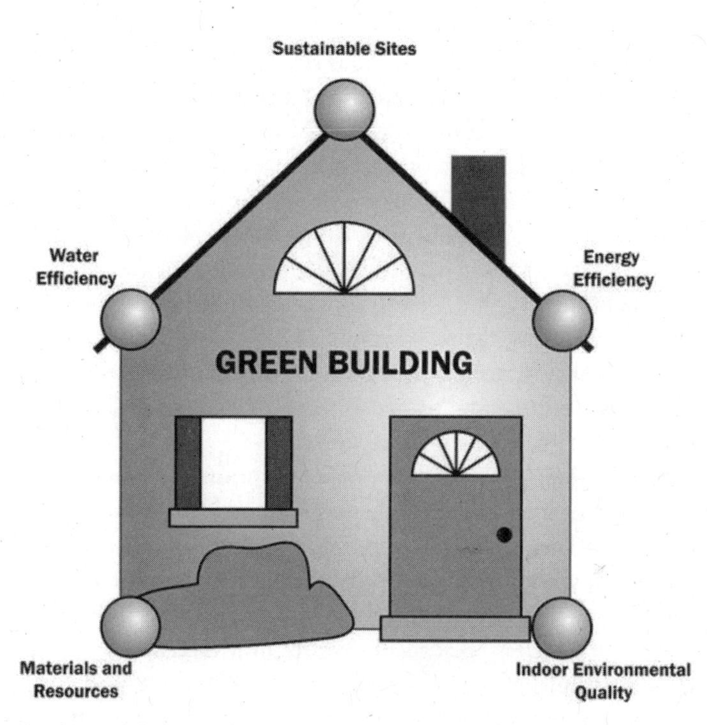

The "house" of green building showing the five major categories of concern.

tive impact of buildings on the environment and occupants in five broad areas:

- Sustainable site planning.
- Safeguarding water and water efficiency.
- Energy efficiency and renewable energy.
- Conservation of materials and resources.
- Indoor environmental quality.[5]

Typically, green buildings are measured against *code* buildings — structures that qualify for a building permit but do not exceed the minimum requirements of the building code for health and safety. In addition, green buildings are often measured according to a system such as the LEED rating system (usgbc.org), the Collaborative for High-Performance Schools (CHPS) ratings (chps.net), the Advanced Building™ guidelines (poweryourdesign.com), Green Guide for Healthcare (GGHC)[6] or, in some cases, local utility or city guidelines (a number of utilities have rating systems for residential buildings). Such buildings must score a minimum number of points above "standard building" performance thresholds to qualify for a certification, or a rating as "green" or high-performance.

Since the introduction of LEED in 2000, it has become essentially the US national standard for commercial and institutional buildings. LEED is primarily a *performance* standard; in other words, it generally allows a developer, architect or building owner to choose how to meet certain benchmark numbers — saving 20% on energy use versus current building codes, for example — without prescribing specific measures. In this way, LEED is a flexible tool for new construction or major renovations in almost all commercial and institutional buildings throughout the US. Canada has an almost identical version of LEED,[7] which has achieved significant popularity. Since its inception, LEED has proven to be a valuable design tool for architectural teams tasked with creating green buildings, as well as a way to evaluate the final result.

LEED provides for four levels of certification, in ascending order of achievement: Certified, Silver, Gold and Platinum. In 2003 and 2004 three projects in southern California achieved the Platinum rating: one project for a local utility, another for a county park (in cooperation with the local Audubon Society) and another for the Natural Resources Defense Council. By early 2007 the largest LEED Platinum project was the Center for Health and Healing at Oregon Health & Science University, in Portland, at 412,000 square feet. At the same time, more than 500 projects had completed the certification process under LEED for New Construction (LEED-NC). Platinum-seeking projects that are under construction in 2007 promise to extend the size of the top-rated buildings to more than one million square feet.

Who Is Using LEED?

By the end of 2006, LEED-NC had captured about 4% to 5% of the total new building market, with nearly 4,000 registered projects encompassing more than 477 million square feet of new and renovated space. At the beginning of 2007, more than 100 new projects each month were registered for evaluation under LEED-NC. Since a project can only be LEED-certified after it is ready for occupancy, many projects are just nearing completion of their documentation to qualify for a LEED rating. Given that it often takes two years or more for projects to move from design to completion (and certification can only take place after substantial completion of a project), growth in the number of certified projects will be rapid. Many Fortune 500 firms, universities, government agencies and non-profit organizations are beginning to participate significantly in the development of LEED projects.

Just about every conceivable project type has been LEED-registered, including a mostly underground Oregon winemaking (barrel-aging) facility! For example, the first 150 LEED Gold project certifications (through the end of 2006) included 10 non-US projects (7 in Canada) and such varied building types as:

- Renovation of a 100-year-old warehouse into a modern office building in Portland, Oregon.
- A developer-driven technology park conversion of an old hospital in Victoria, British Columbia.
- An office-warehouse building for a major auto company in Gresham, Oregon.
- An elementary school in Statesville, North Carolina.
- Two high-rise apartment buildings in New York City.
- A new office building and an office building renovation for Herman Miller, Inc., in Zeeland, Michigan. (Commenting on this project, architect William McDonough observed that moving from a windowless building to a daylit building increased annual revenues 40% and that the increase in profits paid for the building in about four months.)[8]
- A public office building leased to the Commonwealth of Pennsylvania.
- An environmental learning center near Seattle, Washington.
- A city hall in Austin, Texas.
- An affordable housing complex in Santa Monica, California.
- A new convention center in Pittsburgh, Pennsylvania.

CHAPTER 3

What Is a Green Building?

We've been talking about green buildings in general. Now let's get a little more specific about what we actually mean by the term "green building." Utilizing the LEED system of the US Green Building Council, introduced in the previous chapter, a green building is one that is built considering the following five factors. However, most green buildings do not incorporate all of these measures, but rather the project team picks and chooses those that are appropriate for a project's budget and goals.

1. Promote Selection of Appropriate Sites and Environmentally Sustainable Site Development

- Locate projects on sites away from wetlands, above the 100-year flood level, away from prime agricultural land and away from endangered or threatened species habitat.
- Locate projects on sites where there is already urban infrastructure to serve them.
- Locate projects on brownfield sites that have been remediated of contamination; these usually have infrastructure already in place.
- Provide opportunities and building infrastructure for people to commute to work using public transit and bicycles.
- Minimize parking to discourage excessive auto use.
- Provide low-emission vehicles and car-sharing arrangements to reduce gasoline use.
- Protect open space in site development and restore open space on already impacted sites.
- Manage stormwater to reduce the rate and quantity of stormwater runoff, and use best practices to clean stormwater before it leaves the site.

- Manage landscaping and parking lots to reduce excessive areas of open pavement that cause heating of the area around a building in summer, leading to more air-conditioning use.
- Control interior and exterior light from leaving the site, helping to make skies darker at night.

2. Promote Efficient Use of Water Resources

- Control irrigation water use for landscaping, using as little as possible. Select native landscaping which demands little or no added water.
- Look for alternative ways to reduce sewage flows from the project, possibly even treating the wastewater onsite.
- Use water-conserving fixtures inside the building, to reduce overall water demand.

3. Conserve Energy, Use Renewable Energy and Protect Atmospheric Resources

- Reduce the energy use (and environmental impact) of buildings 20% or more below the level of a standard building.
- Use onsite renewable energy to supply a portion of the building's electrical and gas (thermal energy) needs, using solar photovoltaic (PV) panels or solar water heating.
- Commission the building by verifying the functional performance of all energy-using systems after they are installed but before the building is occupied.
- Reduce the use of ozone-harming and global-warming chemicals in building refrigeration and air-conditioning systems.
- Provide a means to troubleshoot the building's energy use on a continuing basis by installing measuring and monitoring devices.
- Supply 35% or more of the building's electrical supply with purchased green power from offsite installations, typically from wind farms.

4. Conserve Building Materials, Reduce Construction Waste and Sensibly Use Natural Resources

- Install permanent locations for recycling bins to encourage the practice in building operations.
- Reuse existing buildings, including interior and exterior materials, to reduce the energy use and environmental impacts associated with producing new building materials.
- Reduce construction waste disposal by 50% or more to cut costs and reduce landfill use.

- Use salvaged and reclaimed building materials such as decorative brick and wood timbers that are still structurally sound.
- Use recycled-content building materials that are made from "down-cycled" materials such as recycled concrete, dry wall, fly ash from coal-fired plants and newspapers.
- Use materials that are harvested and processed in the region, within 500 miles, to cut the transportation impacts associated with bringing them from farther away.
- Use rapidly renewable materials that have a ten-year regeneration time or less, such as bamboo, cork, linoleum, wheatboard or straw-board cabinetry.
- Purchase 50% or more of the wood products in the building from forests certified for sustainable harvesting and good management practices.

5. Protect and Enhance Indoor Environmental Quality

- Provide non-smoking buildings, or separate ventilation systems where smoking is allowed (such as in high-rise housing).
- Monitor delivery of outside air ventilation so that it responds to demand by using sensors for carbon dioxide levels to adjust air flow.

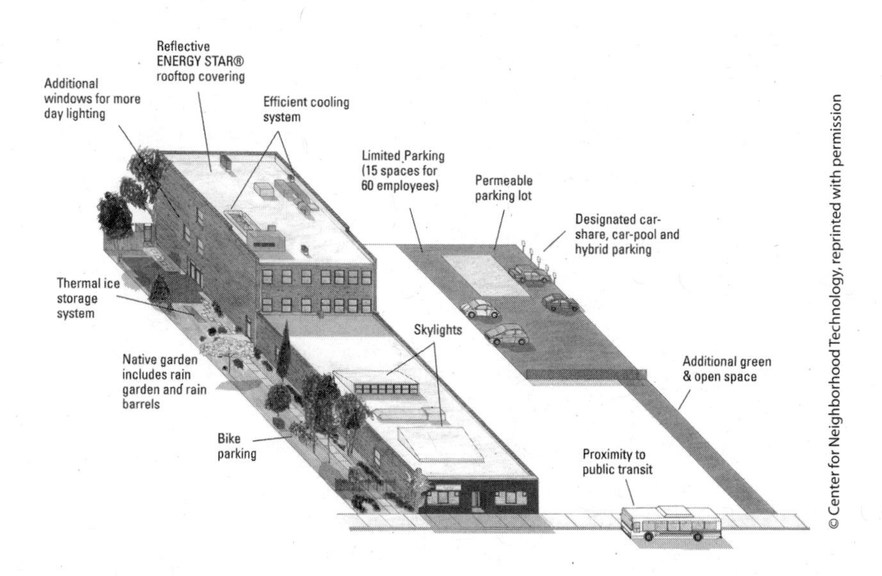

The Center for Neighborhood Technology, Chicago, Illinois, designed by Farr Associates, a LEED Platinum certified project, shows some of the elements that make up a green building project. Energy use is estimated to be 50% less than a standard building.

- Provide for 30% increased ventilation above code levels, or natural ventilation of indoor work areas, to increase the amount of healthy air in the building.
- Conduct construction activities so that there is clean air at the startup of systems and no dust or moisture in materials such as ductwork and sheet rock. The idea is to get rid of "new-building smell" and its associated toxicity.
- Use low-emitting materials in the building to reduce sources of future contamination, including off-gassing from paints and coatings, adhesives and sealants, carpets and backing and composite (or engineered) wood or agrifiber products.
- Make sure that areas where chemicals are mixed or used (such as in-house printing plants or large copy rooms) are separately ventilated, and install walk-off mats or grilles at building entrances to capture contaminants before they enter the building.
- Provide for individual thermal comfort of building occupants, with respect to temperature and humidity.
- Provide for occupant control of building lighting and ventilation systems.
- Provide for adequate daylighting of interior work spaces, using both vision glazing and overhead light sources such as skylights and roof monitors (vertical glazing).
- Provide for views of the outdoors from at least 90% of all workspaces so that people can connect with the environment.

CHAPTER 4

Becoming a Green Building Advocate

In Your Office or Workplace

There are many things you can do where you work to promote green buildings and sustainable design. Here are a few brief suggestions you can implement right away.

Reducing Your Carbon Footprint

In early 2007 Swiss Re, a major global insurance company, announced that it would be supporting investments and purchases made by employees that contribute to reducing carbon dioxide emissions. The new "COYou2 reduce and gain" program is part of Swiss Re's commitments supporting the Clinton Global Initiative. In 2003 Swiss Re declared that it would make its own operations carbon neutral by 2013. Now, as part of the Clinton Global Initiative, Swiss Re has decided to support measures taken by its employees that contribute to the reduction of CO_2 emissions.

The "COYou2 reduce and gain" program supports employees' investments in measures that contribute to reducing greenhouse gas emissions, particularly in relation to mobility, heating and electrical energy. Such measures, which vary according to regional circumstances and preferences, include low-emission hybrid cars, use of public transport and the installation of solar panels or heat pumps. From now until the end of 2011, Swiss Re plans to rebate each employee one-half of the amounts invested in these measures, up to a maximum per employee of 5,000 Swiss francs (about $4,000) or the equivalent in local currency.

According to Ivo Menzinger, Head of Sustainability & Emerging Risk Management, who is in charge of the "COYou2 reduce and gain" program,

"Swiss Re is actively engaged in mitigating climate change and its consequences. This program is an investment that will encourage our employees to make a personal contribution and further raises awareness of the issue."[1]

Take action with your company or business. Some easy steps to take include:

- If you operate a fleet of vehicles, convert them all to hybrids and cut your normal gasoline consumption by 35% to 50%.
- Subsidize employees' use of public transit, at least 50% or more.
- Discourage single occupancy vehicle use by not paying for parking.
- Provide secured bicycle storage in your building with shower facilities or nearby health club passes to encourage people to ride to work in good weather. (This is also a great "wellness" initiative.)
- Buy Green Tags or other "carbon offsets" to cover 100% of your annual travel mileage by car, bus, ferry and airplane. (There are a large number of organizations that cater to this need.)
- Buy green power for the electricity consumption of your workplace; wind-generated power is widely available from a large number of reputable organizations; make sure it is "Green-e" certified from the Center for Resource Solutions.[2]
- Begin the journey to sustainability by examining all of your operations, to see how to reduce their environmental footprint; this activity can involve everyone in the organization; even simple steps like eliminating wastebaskets under individual desks in favor of paper recycling boxes sends a simple message, as does having the IT department set all the printer default setting to "duplex" so people will stop printing on one side of the paper for internal use.
- Undertake a LEED-EB assessment of your existing building operations; LEED for Existing Buildings is a comprehensive evaluation and benchmarking system that will help you "green" your operations and engage the entire workforce in the effort.
- Buy laptops and flat-panel monitors for everyone to cut energy use from "plug loads," often 20% or more of the total energy use of an office.
- Re-lamp and install lighting controls, so you are using only the most efficient fixtures and lights don't operate when people aren't using a room or office.
- Join the US Green Building Council as a corporate or agency member and become part of the solution; once you join, everyone in the company or agency can enjoy the membership benefits.

- Study all of the other aspects of your business operations and work to change each aspect, over time, to more sustainable options, then encourage employees to take those same principles home.

In Your Home or Apartment

The most powerful agent of change is your own personal experience. Think of what you can do to promote green buildings and green operations where you live. Here are a few examples:

- Start keeping track of your gas, electricity and water use, along with the number of gallons of gasoline purchased and airline miles flown.
- Try to cut down on energy and water use by 10% in the next year by examining all of your habits and seeing where you can combine trips or cut down on optional travel.
- Go even beyond 10% reduction: create a "year of living sustainably" that commits you to dramatic changes in lifestyle to meet sustainability goals; if you have kids, enlist their help and creativity. It will strongly supplement the education they're typically getting in school.
- If you can't stop traveling, because of your job or family needs, then start by purchasing "carbon offsets" or Green Tags for all of your mileage, so that you're offsetting the impact with clean power or tree plantings somewhere else.
- Buy a hybrid car or a more fuel-efficient vehicle; you can find the top ten green cars each year listed by the American Council for an Energy-Efficient Economy.[3]
- Look into state and federal incentives for installing solar electric and thermal systems on your home; if you're a renter, discuss the benefits of doing this with your landlord or management company.
- Call the local gas or electric utility company and ask for a home energy audit to find out what are the "low-cost/no-cost" things you can do to cut down on energy consumption; in some areas, the local water company will offer technical assistance or free kits for cutting water consumption.
- Install dual-flush toilets to cut water use from toilet flushing by half or more; install other water-conserving measures such as drip irrigation.
- Form a neighborhood "sustainable living" group to engage the creativity of others in finding additional ways to cut energy and water use, reduce the use of poisons in landscape maintenance and enhance local recycling efforts.
- Consider your purchasing patterns and their "upstream" impacts, including waste in production, transportation costs (if made far from

where you live) and embedded energy of production, distribution, use and disposal.

- For home remodeling, try to support local retail stores that specialize in sustainable products, such as healthy paint and carpet and reclaimed or salvaged building materials.

Your Town, City or State: The Power of Local Initiatives

Just as "all politics is local," a statement famously attributed to former speaker of the US House of Representatives Tip O'Neill, all successful sustainability efforts have their roots in local action. With more than 16 states and 60 cities (as of early 2007) offering local initiatives to promote green buildings, there is ample precedent for you to engage your local school board, city council, county board or commission and even state representatives in this effort. Drill down into each green building success story and you will find just a few local people, some in government, some in business and some plain citizens, whose energy and foresight have made the difference. Some of the initiatives already enacted, on which you can model your efforts, include:

- At the local level, secure a commitment from a school district, city or county to build all future buildings and schools to at least the LEED Silver level; some communities have committed to build LEED Gold projects (the earliest on record was the City of Vancouver, British Columbia); this may take some doing because you're going to hear the old familiar refrain "it costs too much," and you'll have to convince people otherwise by using the examples in this book; among the North American cities making this commitment are Seattle, Sacramento, Portland (OR), Tucson, San Francisco, Calgary and Madison (WI).
- Some cities are taking the next step after greening their own operations, requiring larger private-sector projects to meet LEED certified or Silver-level certifications within the next few years. (Large cities such as Boston and Washington, DC, have done this, and more cities will be requiring such achievements or incorporating LEED requirements and Architecture 2030 milestones into the building code in the next few years.)
- If you have a municipal electric utility or public utility district, convince it to offer incentives for energy conservation and solar energy systems; often the large cash flows of a utility permit it to offer incentives that will, over time, allow it to offset expensive purchases of additional generating capacity in the future; in Texas, Austin Energy, a municipal utility, has been promoting green homes since the early 1990s

Designed by Moore Ruble Yudell Architects & Planners and completed in 2006, the Santa Monica City Library received LEED Gold certification. This building features a central urban location and welcomes the public with lots of daylighting and a central courtyard with its own coffee shop.

and has one of the most successful green home rating systems in the country.

- Convince your mayor or city council to sign onto the US Mayors' Climate Protection Agreement, which commits cities to becoming carbon neutral within the next decade, or sooner, in their own operations;[4] at the global level, former US President Clinton's Climate Change Initiative is engaging the 40 largest cities in the world to become carbon neutral over the next 20 to 30 years.[5] (Already, London has signed on to this initiative.) In Denver, Mayor John Hickenlooper has been aggressively promoting the Greenprint Denver plan for sustainable development,[6] and in Chicago, Mayor Richard Daley has vowed to make Chicago the "greenest city" in North America by promoting green buildings, green roofs and street tree plantings.
- Convince your city council or county commission/board to offer incentives to private sector projects that commit to building green; successful incentives include faster processing of building permits and

increased "density bonuses" for high-rise offices, apartments and condominium developments; if you know a state legislator, talk to them about sponsoring state initiatives to promote green buildings and renewable energy; successful initiatives have included personal and/or corporate income tax credits (Oregon and New York, along with 23 other states); property tax abatements for LEED Silver or better certifications (Nevada); sales tax elimination on solar systems (Arizona, Florida, Georgia, Idaho, Iowa, Massachusetts, Maryland and 12 other states); and rebates for purchase of solar systems (California, Arizona, Colorado and 30 other states).[7]
- Have the governor or state legislature require the state utility commission to have all investor-owned utilities collect a tax on utility bills and offer "public purpose" funds for investments in conservation, onsite power and renewable energy; in 2007 the California Public Utilities Commission adopted an incentive payment system in the form of a consumer rebate, to encourage people to install photovoltaic systems on their roofs; the goal is "a million solar roofs" within ten years.[8]

Your College or University

A college or university is often the largest employer in a town or city; it has a huge impact on energy use, carbon footprint, water use and other municipal services. It also serves as an example to thousands of students, faculty and staff. Make sure that your college or university is doing what it can to promote sustainable operations. A new organization formed to promote campus sustainability efforts, the Association for the Advancement of Sustainability in Higher Education (AASHE), attracted more than 150 campuses as dues-paying members in 2006, its first year of operations.[9] In October 2006 AASHE's first national conference attracted more than 600 people. By early 2007 dozens of campuses all over the country had appointed sustainability directors or coordinators and had begun to implement successful sustainability programs.

One campus that has made sustainability a core part of its mission is Arizona State University (ASU), the country's largest, with more than 60,000 enrolled students, which created, funded and staffed an Office of Sustainability Initiatives in 2005 under the enlightened leadership of President Michael Crow.[10] In 2007 ASU began offering five degree programs in sustainability studies, the first in the country by a major university.

Harvard University has embarked on a major campaign of sustainability initiatives in buildings. The director, Leith Sharp, has run this program since 2001. She reports that the annual return on investment for Harvard's energy efficiency and green building programs is about 36%, about twice that of Harvard's multi-billion-dollar endowment.[11] In other words, to improve their rate of return, Harvard's endowment managers would be well advised to put as much money as possible into the campus's sustainability initiatives! The same could be said for most private universities.

PART II

Green Building: A to Z

CHAPTER 5

Green Building Terms

In this chapter, we present the guts of the book, a brief explanation of 108 of the most important (and up-and-coming) terms used in green building discussions. These terms are typically used by architects, engineers, builders, developers, local officials and building managers to describe the green building attributes of a specific development. Our intention here is not to present a complete description of each topic, but to give you a brief, technically accurate introduction, so that you'll have a better understanding of what people are talking about when the subject of green buildings comes up. At the end of the book, there is a resource section with access to further information, so that you can investigate each topic as much as you please.

Architects and engineers often lapse into techno-speak, using acronyms and terms that even the intelligent and well-informed non-professional can't understand. Stripped of jargon, most green building concepts are understandable to anyone who paid attention in high-school physics and chemistry classes or has ever worked around their own home. Certain terms have been appropriated from general use and have acquired their own specialized meanings, such as "building envelope," a term used to denote the exterior of a building, including the type and amount of glazing (glass) and insulation used.

I have personally trained more than 3,000 building industry professionals through these workshops and given many dozens of speeches and presentations over the past five years. From these, I have garnered an idea of the terminology and concepts that I found initially difficult to understand and that I've seen my audiences struggle with. I've also included some terms that are not in general use but are at the leading edge of sustainability thinking and practice.

So, let's get started!

Architecture 2030

Recent analyses by New Mexico architect Edward Mazria showed that buildings are the source of almost half the global greenhouse gas emissions implicated in widespread climate change. The figure below shows the growth in such emissions up to 2000. Mazria's work and that of the Architecture 2030 organization shows that "stabilizing emissions in this sector and then reversing them to acceptable levels is a key to keeping global warming to approximately 1°C (1.8°F) above today's level."[1]

In 2005 Architecture 2030 teamed with the American Institute of Architects to adopt a policy that "all new buildings, developments and major renovations [should immediately] be designed to meet a fossil fuel energy consumption performance standard of 50% reduction in the 2005 regional (or country) average for that building type."[2] In the winter of 2007, Architecture 2030 sponsored a major teach-in at college campuses to dramatize the issues involved to a new generation of students.

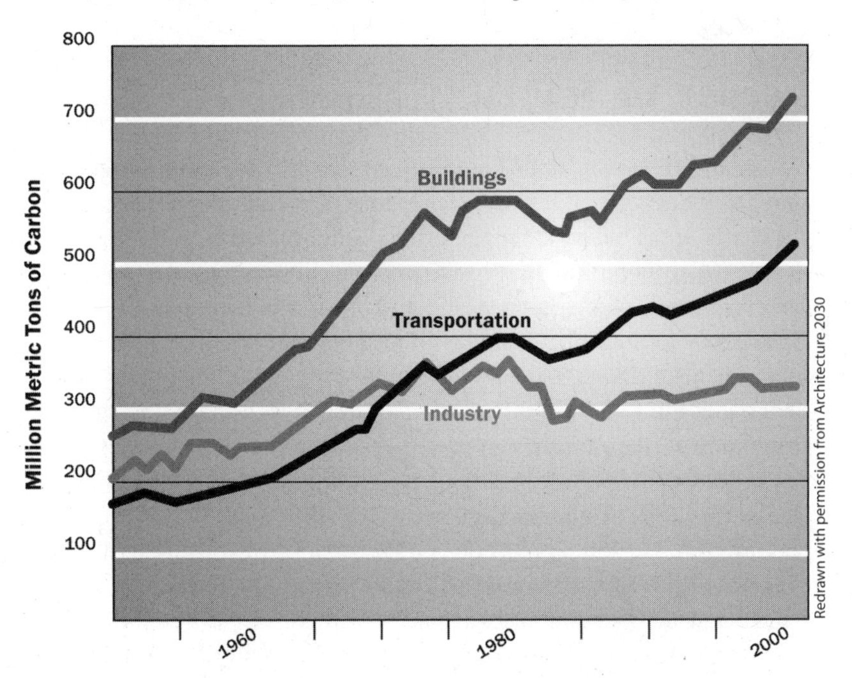

Residential and commercial sector carbon dioxide emissions, 1950–2000. Without changes in present practices, emissions would increase 37% by 2030.

The group further targets a reduction in fossil fuel use in new buildings to achieve reductions (against the 2005 average energy use of commercial buildings) of the following magnitude:

- 60% in 2010
- 70% in 2015
- 80% in 2020
- 90% in 2025

According to this proposal, by 2030 all new buildings should operate as carbon neutral, that is, requiring no emission of greenhouse gases, either directly or indirectly, to operate. Most architects would tell us that, at little or no additional cost, most buildings could dramatically lower their energy consumption through proper orientation, building shape, selection and placement of glazing and by incorporating natural heating, cooling, ventilation and daylighting strategies. Additional energy needed to make a building work could then be supplied by renewable energy sources and by biofuels.

Access to Transit

One of the great virtues of the LEED green building rating system (and similar systems) is that it encourages projects to employ site selection criteria early in the process that go beyond economic issues and instead deal with broader environmental concerns. Since automobile travel is one of the largest users of energy in the US and one of the major contributors to urban smog formation, it makes sense that commercial projects should locate near mass transit, so that workers have the option of using it, rather than being forced to drive. In many cities, such as Portland, Oregon, the cost of downtown parking now exceeds $150 per month, a strong incentive not to drive a car to work.

In the LEED system, projects are encouraged to locate within a quarter-mile of two or more public or campus bus lines or within a half-mile of a funded or planned commuter rail, light-rail or subway station. These are walkable distances in most cities (even in hot or cold or inclement weather) and are feasible locations in many older suburbs. In newer transit-oriented developments, access to transit is a key marketing feature of

The Swinerton Inc. headquarters in San Francisco is a LEED-EB Gold-certified building that provides excellent access to rail, bus and ferry transit services for its employees.

office buildings. Companies are supporting transit use by employees through subsidized monthly bus or rail passes.

A transit-oriented development (TOD) is a residential or commercial area designed to maximize access to public transport and often incorporates features to encourage transit ridership. A TOD neighborhood typically has a center with a train, light-rail or bus station, surrounded by relatively high-density development, with progressively lower-density development spreading outwards from the center. This type of arrangement contains specific features that are designed to encourage public transport use, including mixed-use (residential, retail, office) development that will use transit at all times of day, high-quality pedestrian crossings, narrow streets and tapering of building heights as they become more distant from the public transport node. Another key feature of transit-oriented development is the reduced need for parking spaces or additional parking garages.

b

Bicycle Commuting

In addition to giving access to public transportation, green buildings very often provide accommodations for bicycle commuters, since the cost of doing so is typically small relative to overall project costs. In many cities that are bicycle-oriented, increasing numbers of people are commuting to work in this healthy, friendly, non-polluting way.[3] In my former home of Portland, buses and light-rail cars were outfitted to hold bicycles for the long trek home, especially in darkness (evening commutes in the winter) and rainy weather. Madison, Wisconsin, has the most mileage of bike trails for a city of its size. Over the past 30 years, the city has instituted more than 30 miles of bike paths and 110 miles of on-road bike lanes. For trips under five miles, it's faster to pedal than drive in many cases.

In addition to the obvious benefits of cutting gasoline use and reducing air pollution from automobiles, bicycle commuting can (over the long haul) reduce traffic congestion, noise (except for bicycle riders screaming curses at inattentive drivers) and lower infrastructure investment for parking lots. Generally, bicycles are used for shorter commuting trips, which help cut emissions from vehicle startup. There are clear health benefits from bicycle commuting that are important in this era of growing concern over the nation's epidemic of obesity.

To encourage bicycle commuting, the key for building designers is to provide showers (one shower for every ten users), changing rooms and safe bicycle storage. Showers are also a building amenity, since they can serve anyone who runs or otherwise exercises during a lunch break at work. The LEED standard requires bicycle racks or secure storage for 5% of a building's full-time peak occupants, either onsite or within 200 yards (such as a campus environment). So, if a building has 400 people, bicycle racks or storage lockers for 20 people would be required, along with two showers (these can be unisex, with locking doors on both sides).

Designing for bicycle commuting is neither expensive nor difficult, so it should be standard in every office building, shopping mall, campus building and other urban institutional and commercial uses. It is also socially responsible, since it doesn't force anyone to own a car to get to work, as long as they are able to ride. Combined with a car-sharing program that allows people to rent a car by the hour for emergencies, sustainable design offers easy solutions to enhance a gasoline-free lifestyle. (For residential buildings, the LEED standard requires covered bicycle storage for 15% of building occupants, in lieu of showers and changing facilities that are obviously not needed.)

● ● ●

Big Picture

As our discussion of the Architecture 2030 program shows, there is a big picture of global climate change, species loss, rainforest destruction, soil erosion and a myriad of other environmental issues on the table, as human beings attempt to accommodate an estimated 9.4 billion people on the planet by 2050 (up from about 6.6 billion today) at reasonable levels of health and material well-being.[4]

The big picture was dramatized to many of us, in 1969 by the first pictures of the entire Earth from outer space, taken during moon orbit by NASA astronauts, showing one planet, blue and alive, in a sea of emptiness, black and hostile to life. The picture of "Earthrise," taken as a circling lunar module emerged from the dark side of the moon, was particularly dramatic, inspiring and life-changing for many. We do have to go it alone on this planet, science fiction fantasies of future space colonization notwithstanding. As human activity diminishes the productivity and diversity of natural systems, a process well underway all over the world, we increasingly face the prospect of creating a new inhabitable Earth in our own image, without the requisite knowledge or humility.

As an example of the big picture, consider the Gaia Hypothesis, a scientific theory that proposes that the living matter of planet Earth functions like a single organism. It was first formulated in the 1960s by the independent research scientist James Lovelock, as a consequence of his work for NASA on methods of detecting life on Mars. The Gaia Hypothesis, now known as "earth system science," has since been supported by a number of scientific experiments and provided a number of useful predictions.[5]

Another example of the big picture addressed by green buildings, with their emphasis on dramatic reductions in energy use and global greenhouse gas emissions, is the shrinking of the Arctic ice cap by 20% since 1970 and the visible shrinkage of northern hemisphere glaciers in Europe and America from global warming.[6] In Portland, contemporary photos of the nearby Mt. Hood glacier show dramatic shrinkage since the first aerial photos taken in the 1920s and 1930s. Many people have now seen photos and drawings showing the significant reduction in Arctic ice over the past 30 years. This reduction leads to fewer ice floes for polar bears, leaving them with a choice of returning to land for much of the year or drowning because of the vast space between ice floes. Some hypothesize that the melting of Arctic ice could trigger a runaway climate warming because so much of the sun's energy during the summer would be absorbed instead of being reflected back into space. Fresh water would stream into the North Atlantic, upsetting the circulation pattern of the Gulf Stream that currently makes high latitudes in Europe suitable for agriculture and high-population densities.

● ● ●

Biodiesel

Biodiesel is diesel-equivalent, processed fuel derived from biological sources (such as vegetable oils, even waste cooking oil from restaurants such as fast-food establishments), that can be used unmodified in diesel engines. Biodiesel is biodegradable and non-toxic and produces significantly fewer emissions than petroleum-based diesel when burned. A recent article reported that chicken fat could be used, prompting major chicken producers such as Tyson Foods, Perdue Farms and Smithfield Foods to set up renewable energy divisions to sell the material mixed with soybean oil.[7] It's possible that, sometime soon, the tag line for biodiesel may change from "smells like French fries" to "smells like fried chicken."

One benefit of using biodiesel is said to be lower engine wear. Most manufacturers release lists of the cars that will run on biodiesel. For example, Volkswagen determined that diesel fuel containing up to 5% biodiesel (B5 fuel) meets the technical specifications for its vehicles equipped with TDI engines in the United States.[8] Biodiesel can be distributed using today's infrastructure. As a result, its use and production are increasing rapidly. For example, at the end of 2006 there were 635 biodiesel fueling stations in 46 of the 50 states, with South Carolina, Missouri, North Carolina and Texas having the most outlets. Early in 2007 the Safeway grocery chain began selling biodiesel at a store in Seattle, Washington, with plans to expand distribution if the market test proves successful.[9]

Just as fuel stations begin to make biodiesel available to consumers, a growing number of trucking fleets use it as an additive in their fuel. According to the National Biodiesel Board, US biodiesel production tripled in 2005 to 75 million gallons from 25 million in 2004, with 2006 numbers expected to increase to between 150 and 225 million gallons. One expert predicted that by 2012 it will exceed one billion gallons per year.[10] This is obviously a small contribution to reducing petroleum dependence for transportation, but worth doing in any case because it converts waste material with high fuel value into a useful product.

For green builders, biodiesel can be used in construction vehicles and in diesel fleets maintained by large retail and manufacturing companies professing sustainability as a value. Consider the value of greening your operations with a focus on biodiesel fuel. For example, construction firms self-performing concrete work can convert their concrete trucks to run on biodiesel and also use it in their fleets of pickup trucks and other vehicles.

What can you do? Consider asking your firm, agency or institution to subsidize biodiesel conversions or fuel purchases to kick-start this new industry in your town or city. In this way you will be taking action to assure a more sustainable future, one less dependent imported oil.

● ● ●

Biophilia

The Biophilia Hypothesis postulates that human beings have an instinctive bond with all other living things, a theory first proposed in the 1980s by biologist Edward O. Wilson.[11] Designers are increasingly recognizing the importance of connecting people with the outdoors through building design, by bringing nature into buildings and buildings into nature.

For example, in green buildings, great importance is placed on providing views of the outdoors from all workstations. Research indicates that being able to see outside during the workday is more conducive to physical and mental health than working in a windowless environment.

We shouldn't be too surprised at such a result. After all, just recall that human beings evolved for the past two million years in intimate connection to the natural world, depending on it for food, shelter, clothing and tools for survival, as well as for poetic and artistic inspiration. Most indigenous cultures are intimately related to their specific place on the planet; the animals, birds, vegetation, and creatures of the sea and rivers occupy a special place in their creation stories and sense of well-being. Only during the past 100 to 150 years have a large number of people spent most of their daylight hours indoors; it seems that we are still hard-wired to want to know what's going on around us.

Daylighting design is also a way to bring nature and the natural cycle of the sun to our attention. Research on 21,000 elementary school students in 1999 showed that schools with daylighting and views of the outdoors promote higher test scores among students; the more windows and views, the higher the score.[12] Studies of office workers in California in 2003 concluded that "better access to views consistently predicted better performance."[13] Daylighting has also been found to increase retail sales at chain stores by 5% to 40%, with a profit value 20 to 50 times the value of just the increased energy savings from skylights.[14]

Green building designers have gone much further to explore biophilia. For example, past studies of healing in hospitals have shown that patients heal faster (and get out of the hospital faster) when their windows face onto natural areas. Many hospitals have taken this idea into the form of healing gardens where patients recovering from surgery can spend time each day.

Many of us have spent time in buildings where there are waterfalls, winter gardens, even simple natural elements such as boulders and sand, and we've noticed how much more relaxed and comfortable (and comforting) such places are. Isn't it amazing that we put so many people to work each day in sterile cubicle farms and then expect them to be productive? How little knowledge we really have of human motivation, health and performance in the world of commercial architecture and design! Biophilia promises to be an interesting, exciting and vital part of green building design over the next decade.

Brownfields

Most urban areas have a large number of former manufacturing, warehousing and transportation sites that have (or have had) varying degrees of pollution, ranging from petroleum-contaminated soil to places with significant pollution from lead, mercury and various heavy metals. Most of these sites can be remediated and restored to productive use. From a social viewpoint, they also typically have roads, water supply, sewerage and storm drains, valuable infrastructure that can support extensive new de-

Atlantic Station, Atlanta, Georgia.

velopment without forcing the public (or a developer) to bear a lot of new investment.

This situation was brought home to me during a visit to Hawaii in the summer of 2006, when I attended a conference at the new Honolulu Convention Center near Waikiki. About 15 years earlier, I had directed an environmental remediation project at the convention center site, hauling some 3,000 tons of petroleum-contaminated soil by barge back across the Pacific Ocean to Washington state, to be deposited in a new state-of-the-art landfill east of the Cascade Range. (There had been a car dealership and some public vehicle maintenance facilities on the site in earlier years, the likely source of contamination.) Having stood vacant because of industrial contamination for many years, the land was back in productive use as a high-profile urban area.

In 2005 I visited Atlantic Station, a $3 billion real estate development in the mid-town area of Atlanta, on 138 acres at the site of the former Atlantic Steel manufacturing plant. The residual contamination had been cleaned up and in its place was a thriving urban village of homes, shops and offices. The developers took a green approach to the overall development, starting with site remediation, and the value of the real estate was many times what it had been.[15] At this project, the 171-17th Street office tower received a LEED for Core and Shell certification at the Silver level, the first such designation for a high-rise building in the country. For the developers, AIG Global Real Estate and Jacoby Development, the publicity value was enormous. The Atlantic Station project combines green buildings with smart growth redevelopment strategies, putting a mixed-use development close to a light-rail station.

Building Codes

A friend in Tucson, David Eisenberg, in addition to being a national expert on straw-bale housing construction technology,[16] has been heavily involved with building codes and building code officials the past half-decade or more, as his way to change the environment to favor sustainable design. David says:

> I start with this premise: Building codes are based on a societal decision that it is important to protect the health and safety of people from the built environment. If, inadvertently, the codes are actually jeopardizing the health and safety of everyone on the planet by ignoring their impacts on resources and the environment, resulting in the destruction of the ecosystems that sustain us, we are obligated to reinvent the codes with that larger perspective.[17]

For building designers, it's often necessary to challenge the building codes to make advances that both save energy and save money in construction. In designing a LEED Platinum healthcare building, Interface Engineering in Portland secured 11 successful code appeals, effectively saving the project hundreds of thousands of dollars while improving the energy efficiency and rationality of the overall design.[18] As one example, the engineers figured out that the underground garage ventilation requirements in the code were based on outmoded data on vehicle emissions of carbon monoxide (the odorless, colorless, toxic gas that will kill you if there's too much of it). By finding and submitting up-to-date information on actual emissions of a typical vehicle fleet, the engineers were able to reduce the fan size by 60% and its energy use by 60% for the life of the building. In addition, they set a precedent for all future projects in the city, thereby saving money for other projects with underground garages and saving all the energy from oversized fans moving too much air for the life of each building.

Scratch the surface of most codes and you'll find similar money-saving and energy-saving opportunities. The trick is to convince architects, engineers, project managers and building owners or developers to push the envelope in each code jurisdiction and with each project, so that local building officials get comfortable with new technologies. Some green techniques that may require appeals and variances include onsite graywater use, onsite sewage treatment and reuse, water-free urinals, underfloor air distribution systems, constructed wetlands for stormwater management and wastewater treatment, natural ventilation systems and innovative approaches to fire and life-safety protection.

Code officials are generally like everyone else: they want to do a good job, they are risk-averse and they need to trust the people they're dealing with. I have found that when local engineers and architects are established in a given town, it's generally easier to get code officials to listen to new approaches: they know they can rely on the professional judgment and experience of the designers, and they know they'll be around to fix problems if something doesn't work.

● ● ●

Carbon Dioxide Monitoring

Imagine you are in a meeting room in the afternoon. After 20 people are there for a while, the room gets stuffy, hot and generally uncomfortable. Sound familiar? It's because the meeting room has a constant-volume air supply, rather than a variable-volume supply that would increase ventilation when the room is occupied by lots of people. Sound simple? It is, and that's why most green building projects use carbon dioxide monitors to regulate ventilation levels in workspaces, with feedback to the building automation system that controls the fans and fresh-air intake.

In most buildings, carbon dioxide levels build up during the day, causing us to become sleepy after lunch in many cases. This hampers productivity and reduces comfort. It's like being on a warm, humid tropical beach, quite enjoyable during a vacation but not really conducive to regular office tasks. Typically, discomfort is triggered by carbon dioxide levels that are a few hundred parts per million above naturally occurring levels in the atmosphere.

Carbon dioxide monitors work on a simple precept: we breathe out carbon dioxide and take in oxygen; by monitoring carbon dioxide levels and using known levels of ventilation, we indirectly know how many people are in a room. Then we can adjust ventilation so that there are always a certain number of cubic feet per minute of fresh air per person. The monitors are relatively cheap and should be used in every project. The only issue is for engineers to specify good-quality monitors that hold their calibrations for several years, so that the ventilation levels remain appropriate for occupancy levels. The calibration needs to reference the baseline levels of carbon dioxide and to set limits for indoor air, so that there is always sufficient ventilation to keep the building air fresh.

The LEED standard for new construction handles this issue by requiring carbon dioxide monitors for densely occupied spaces (less than 40 square feet per person) and direct outdoor airflow measurements for regular office spaces, to achieve at least design-minimum ventilation. For naturally ventilated spaces, LEED requires carbon dioxide monitors between three and six feet above the floor.

Since most naturally ventilated spaces are typically "mixed mode" with some fan assist for stagnant air days, the monitors can be used to trigger the fans to pull outside air into the building. In buildings with no fans but with operable windows, a good green building design would have an alarm or signal to tell people to open the windows to let in more outside air.

Carbon Neutral

Most organizations interested in sustainability are beginning to put together programs to be carbon neutral in their activities. A West Coast building engineering firm (and a former employer of mine) with about 200 employees sent out a 2007 New Year's greeting with a carbon neutral announcement:

> In 2006, with the help of our sustainability partners American Forests, Glumac has achieved its New Year's resolution of becoming a Carbon Neutral company. In the last year, as part of our ongoing commitment to the environment, 15,000 trees were planted in forests throughout Oregon and California. These trees will offset the entire amount of carbon dioxide emissions generated by all of our offices, our projects and employees.[19]

In addition, to further its commitment to being carbon neutral, the company says that it will:

- Purchase and operate hybrid vehicles for company cars.
- Subsidize public transportation for all employees.
- Use Flexcar (rentals by the hour) for local meetings and short trips.
- Green the company's offices in Portland and San Francisco with Silver LEED Certification.
- Build a new office in California to achieve Platinum LEED Certification.

Every company, non-profit and government agency can take similar steps to become carbon neutral in their operations, as can design firms in their projects' environmental footprints. Larger organizations can buy carbon offsets for their activities from a variety of free-market carbon exchanges, by financing activities that reduce the production of carbon dioxide and other greenhouse gases such as HFC-23, a by-product of HFC-22 refrigerant production. According to the World Bank, the carbon-credit market was valued at $21.5 billion for the first three quarters of 2006, about double the value for all of 2005.[20] The biggest source of carbon credits at present is China, the second-largest energy user on the planet, with a lot of inefficient older power and chemical plants. Under the current rules, only chemical factories operating prior to 2001 qualify to sell carbon credits for the destruction of HFC-23.

Going carbon neutral may still be problematic for some, owing to expressed concerns over whether all the money being spent actually goes to a useful purpose, but it is gaining currency as the next logical step for individuals and companies to take action against global warming by offsetting their carbon dioxide emissions.[21]

Carpet

Everyone knows new carpet smell, because it lingers for weeks or even months after carpeting is installed. I had that experience recently in recarpeting part of my home in Tucson. Despite my efforts to find *low-VOC* (volatile organic compounds) carpets from local retailers, I was unsuccessful, and I just had to live with the smell until it dissipated. This drove my wife nuts because she has multiple chemical sensitivities and is consequently very allergic to any chemical odors. So we just opened the windows and aired out the house for a long time.

In the commercial arena, however, the situation is quite different. Most major manufacturers of commercial carpeting have agreed to meet the LEED standard for Green Label Plus, as defined by the testing and product requirements of the Carpet and Rug Institute. Carpet backing has to meet the Green Label standard. What does this mean? Technically, it means that the off-gassing of VOCs from carpet adhesive, for example, is limited to less than 50 grams per liter, under standard test conditions. Practically, it means an end to the new carpet smell that bothers a significant minority of people in offices, libraries, public facilities and schools.

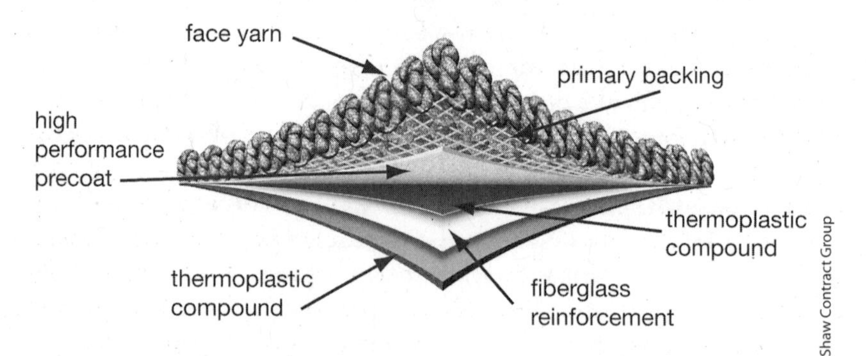

Shaw Contract Group's new EcoWorx Performance Broadloom Carpet.

The good news is that most of the major carpet manufacturers support these standards, at least for the commercial market. They are quite competitive; no one wants to have another company get an edge in the marketplace. In this situation the private sector is working as it should — innovation is rapid, new products are rolled out every year and the ecological footprint of the industry is lower than it would be otherwise.

Broadloom carpet makes up about 70% of the commercial flooring market. In 2007 Shaw Contract Group launched *EcoWorx Performance Broadloom*, claimed to be the commercial carpet industry's first fully sustainable cradle-to-cradle carpet backing for broadloom. The product incorporates components designed to maximize performance and recyclability. Because of this approach, the carpet can be easily generated, deconstructed and regenerated through a five-step dissolution process.[22]

Certification

One of the most common questions I hear asked about green buildings is: If we're already doing the "right thing," why should we bother to certify a project and incur extra costs? The basic answer is: If you don't go through the documentation and certification process, how do you know what you actually did? Certification provides a recognized third-party verification of achievement. It may surprise people outside the design and construction industry, but when a building is finished and occupied, almost no one has a definitive idea of what went into the building and if all the systems are actually going to work! Imagine a $30 million movie production going forward without someone finally responsible for what goes on the screen.

The process of certification starts with the initial design meetings, where the goals of a project are reviewed and mapped against the LEED evaluation system. When design is finished and a project goes out for contractor bid, then the design aspects of the project can be reviewed by the USGBC, and the certification process begins in earnest, as the architects and engineers begin filling out the LEED information templates.

Certification has multiple values but is essentially a quality assurance and performance verification process. Recently someone called me about a public project in Alaska. Their building committee chair said he was sure that they had done a good job of sustainable design and didn't see why spending the money for a formal certification was necessary. When I heard this, my response was: Sustainable design, you say, but against which metrics? How do you know it's a green building if it doesn't measure up to an accepted standard such as LEED? How will you prove to a skeptical public it's a green building if no one has a clear definition of what this term means, or no responsible third-party attests to the achievement?

If you're a public agency, corporation, developer or non-profit, you likely have stakeholders who care deeply about your commitment to sustainability and reducing the carbon footprint of your buildings. They want to know that you're doing the right thing, and obtaining LEED certification of your new building is one way to convince them of it. Otherwise, your claims are suspect. The likelihood of being accused of greenwashing is much smaller, and the credibility of the claims for greenness is much more believable if a project is certified. From a marketing and public relations point of view, third-party certification by a recognized organization such as the USGBC has enormous credibility with the public and the press.

Certifications can also trigger tax benefits in a number of states, and the LEED-required energy-use modeling and building commissioning have immediate payoffs in evaluating design decisions and in ensuring that all energy-using systems actually talk to each other when the building is operating. Certification has a cost, typically $50,000 to $100,000 or more, if you count energy modeling and building commissioning, which should be done for any quality building. For a really small building, that cost can be burdensome, but for larger projects it's well within the contingency budget.

● ● ●

Certified Wood Products

Certified wood products are those made from lumber harvested in a sustainable manner and certified by a reliable third party. The certifying groups most active at this time are the Forest Stewardship Council (FSC), Sustainable Forestry Initiative (SFI) and the Canadian Standards Association (CSA). The LEED rating system only awards points for FSC-certified wood, partly for historical reasons and partly because it is the most rigorous third-party rating system. However, Green Globes and the National Association of Homebuilders (NAHB) Green Home Building Guidelines also recognize the SFI and CSA systems. As a practical matter, most of the timber harvested in the US comes from public lands and is not certified to

© 2003 The Collins Companies, reprinted with permission

Bainbridge Island, Washington, City Hall, features FSC-certified CollinsWood hem-fir and ponderosa pine from Collins Pine Company's forest in Chester, California.

any standard. Among certifications, typically from private woodlands, FSC has about 25% of the market (at the beginning of 2007) and SFI about 75%.[23]

Certified wood products all carry a "chain of custody" certificate that tracks the lumber from the forest to the end-user. LEED-certified projects must use certified wood products for 50% of the value of all permanent wood-based materials in a building, including flooring, dimensional lumber (2-by-4s and the like), subflooring, roof decks, paneling, door cores and cabinetry. Softwood, typically used for structural purposes, carries little cost premium, while hardwood, used in finished carpentry such as cabinets and furniture, still carries some cost premium, often depending on whether the seller is vertically integrated (owns their own woodlands, mills, etc.) or not.

One very interesting development in the past few years has been the growth of underwater salvage logging from thousands of reservoirs around the world that have dead but usable trees. Think of a reservoir as a drowned valley, and you'll appreciate the beauty of the concept. A Canadian company, Triton Logging, has developed a unique, remote-controlled submersible logging machine (dubbed the Sawfish™) to do the trick and to float the logs to the surface. Triton Logging expects its SmartWood Rediscovered certified (by the Rainforest Alliance) salvage lumber to be widely used in green building projects, since it can be harvested about 20% cheaper, with a much smaller environmental footprint, than standing timber in a forest.

Charrettes and Eco-charrettes

Charrette is French for a "small cart"; a charrette was used to transport people to the guillotine during the French Revolution; later in the 19th century, architecture students at famed L'École des Beaux-Arts in Paris rode in horse-drawn charrettes to their final examinations, clutching their drawings, all finished at the last minute. From this inauspicious beginning, architects began to treat a charrette as an intensive design exercise, in which project participants work together for a day or more until a design has been worked out or at least until areas for further study are clearly assigned to each participant.

To my knowledge, the so-called eco-charrette was first named in the late 1990s by Nathan Good, an architect in Salem, Oregon, to denote a focus on the sustainable design aspects of a project. Charrettes are facilitated sessions that utilize the skills of all participants to arrive quickly at major design decisions, with full recognition of all the potential interactions of green building measures with building requirements. The purpose of an eco-charrette is to explore the key green building and green development aspects of a project *before* any important design decisions are set in stone. Typically it occurs early in the schematic design process (see the section on Integrated Design).

In my own experience, eco-charrettes can help discover unexpected synergies between disparate design items. In one project, the engineers decided to ask if they could put a radiant heating system in the concrete floor slab of a large atrium. As a result, they were able to provide both supple-

Green Building Services, Inc.

Consultant Ralph Dinola of Green Building Services, Portland, Oregon, conducts an eco-charrette.

mental heating and cooling from the tubing placed in the floor near the top of the concrete. The additional cost of added concrete and tubing was minor, far less than the money saved by reducing the HVAC system size.

Along with an eco-charrette, which often involves fairly detailed and technical discussions of design alternatives, I have often found it useful to have a visioning or goal-setting session with key decision-makers who will not be involved with the more technical aspects of the project. These sessions should involve representatives from the occupants of the building, along with "C-level" executives (CEO, COO and CFO) who will have to approve the expenditures for the green building certification.

Comfort

Human comfort plays a major role in productivity. If people are too hot or too cold, some of the mental energy that should go to productive work gets diverted into figuring out how to cool off or warm up. The primary environmental determinants of human comfort include air temperature, radiant surface temperature, air velocity and relative humidity. Ultimately, comfort is defined by LEED as "a condition of mind experienced by building occupants expressing satisfaction with the thermal environment."[24]

From an engineer's perspective, the type of clothing worn also influences comfort. Wear a wool suit on a summer day, and you just won't feel as comfortable in the air-conditioned office as you would with a short-sleeved shirt. And, of course, men and women, young and old, thin and fat, all feel differently in terms of comfort. How much you move around on the job also plays a role; sitting in front of a computer all day will likely make you more sensitive to temperature swings.

Standards of comfort as specified by the American Society for Heating, Refrigeration and Air-conditioning Engineers (ASHRAE) include consideration of both temperature and relative humidity. Most mechanical engineers learn how to design ventilation and space-conditioning systems using ASHRAE methods and guidelines. Since 2005 ASHRAE, AIA and the USGBC have teamed up to begin integrating their standards to reduce energy use and improve comfort in buildings.

The LEED rating system recognizes that human comfort plays a significant role in productivity and health in buildings, so it gives a point for maintaining established comfort standards throughout the year. In some colder northern climates, buildings need to add humidity during winter to gain comfort, since outdoor cold air tends to be very dry. In hot, humid climates such as the Southeast US, moisture needs to be taken out of the air almost year-round since high humidity even in mild weather can be uncomfortable in closed spaces. Warm summer afternoons can be particularly challenging for air conditioning if the architectural design doesn't provide external shading on south- and west-facing walls and windows. Then it's almost impossible to prevent a building from overheating in the late afternoon.

Natural ventilation systems are another approach to providing comfort and fresh air. If a building is designed to circulate air naturally from outdoors to indoors, and then to provide a "stack-effect" exiting path, where the heated air, rising naturally, exits the building near the roof, sometimes through an atrium or stairwell, it can be comfortable even if temperatures are a bit cooler or hotter than normal. Think of your own experience on a warm spring day when you can feel the outdoor air moving through the building. The sensory experience of the slowly moving air, fresh from outdoors, overwhelms any feelings of discomfort because of slightly higher or lower temperature than normal.

Commissioning

Building commissioning is a high-value-added activity that is unknown outside the building industry. Think of a ship; when construction is finished, it's time for sea trials. Long before a vessel sets out on a mission or voyage, all key systems are tested in calmer waters to make sure everything is working as designed. This includes propulsion, navigation and safety equipment.

Now consider a modern high-rise building, which is every bit as complex as a ship. It's expected to perform well for decades, supporting all types of occupancy and enduring both normal and extreme weather events, including torrential rains, high winds, tornadoes, floods and hurricanes, and to be safe for its occupants in the event of fire or earthquake. Shouldn't this building be commissioned just as a ship would be?

Posed this way, the answer is obvious. In the past decade, the practice of commissioning for larger buildings has become an accepted practice. The LEED system requires that every project be commissioned according to certain standard procedures. The goal is to test all energy-using and life-safety systems in actual building operation and to work out all the kinks before occupancy. More than 120 research studies have shown that energy savings increase 10% to 15% when a building is commissioned.[25] In energy savings alone, commissioning pays for itself in less than five years; when other non-monetary (but real) benefits are included, the return is typically less than one year.

The cost of commissioning is relatively minor compared with the benefits. In larger projects, the cost might range from $0.40 to $1.00 per square foot, less than 1% of building costs. The key to the process is to get experienced commissioning agents on board during the design phase so that they can understand and help clarify the owner's project requirements and the engineer's basis of design. In this way the commissioning agent understands the project's goals, systems and performance requirements before testing begins.

A typical commissioning activity involves creating a plan; writing commissioning requirements into the project specifications; engaging the subcontractors during construction — especially mechanical, electrical and controls contractors — to assist with testing; fixing any problems encountered with system operations; and confirming that operators have been trained to keep the building running optimally.

Some projects also provide enhanced commissioning, including having a completely independent commissioning agent reporting directly to the building owner (some people think this provides a more objective look at the project functioning), examining design documents well before construction starts, and coming back near the end of the warranty period (typically a year) to examine the building performance and correct any deficiencies. Enhanced commissioning also requires someone to prepare a re-commissioning manual for later use, since commissioning is best regarded as a continuous activity over the lifetime of a building.

● ● ●

Controllability of Systems

Another aspect of comfort is the degree of control you have over your surroundings, particularly lighting, air flow and temperature. Underfloor air systems combined with task lighting tend to provide the highest levels of control, and there are even systems that take advantage of this fact to craft personal environments that allow users to control ambient noise levels, along with these three variables.

Buildings with operable windows can also provide user control, which heightens perceived comfort. Mechanical engineers tend to strongly dislike operable windows, because they make it difficult to control air pressures and require more zones on each floor. Of course, building HVAC systems need to be shut off in zones where the windows are open. In tall buildings, a strong wind can come in the upper-floor windows, so occupants must keep papers and other belongings from blowing around. In North America we are just beginning to see operable windows. In many European countries accustomed to letting nature provide the comfort, they are commonplace.

In 2002 I visited a new green building, the Telenor headquarters in Oslo, owned by Norway's state-owned telephone company. There employees work in 30-person task groups under one manager. When I asked who determines when to open the windows, the tour leader answered, "The manager." In the US I'd wager it would be easier and less contentious to have these decisions made impersonally by the building automation, instead of by a manager.

An early LEED Platinum-rated building, the Chesapeake Bay Foundation offices in Annapolis, Maryland, provided one way to solve this prob-

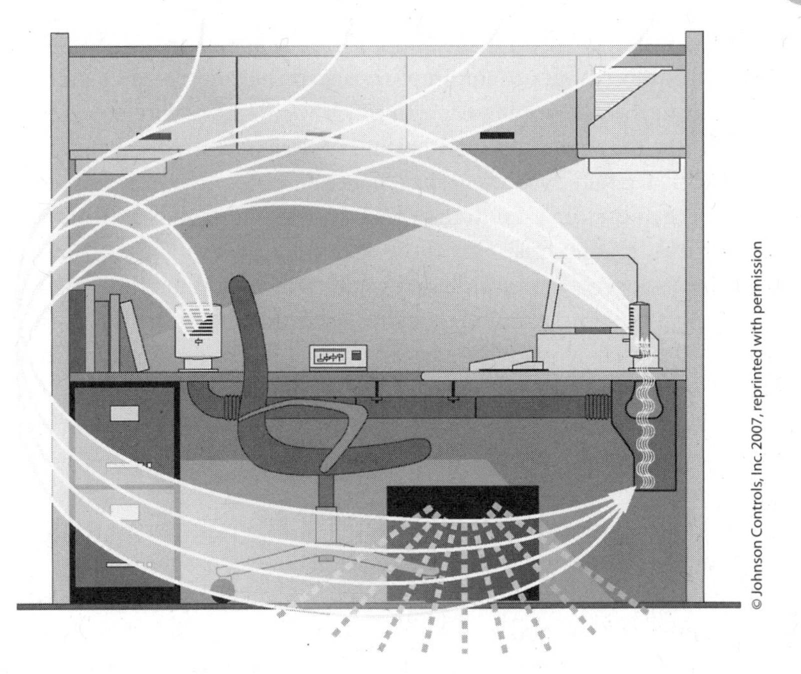

Johnson Controls Personal Environments allow user control of key comfort variables.

lem. When I visited this building in 2001, I was curious about the operable windows. I found that the designers had made the decision impersonal: whenever temperature and humidity conditions were favorable, a green light came on, and the 40 or so occupants rushed to the windows to open them. The Foundation was an environmental nonprofit, so you might think that they wanted to be more involved with the building and more in touch with the outdoors, but I don't think they are at all unique. A developer in Portland told me that a prime selling point for a new LEED-certified office space, leased to a major law firm, was that it had operable windows.

Construction Practices

With all the focus on the role of architects and engineers in creating sustainable designs, it's easy to overlook the vital role played by the construction team in green building achievements. Contractors have the expertise to translate designs into finished buildings; often, they are instrumental in suggesting better ways to accomplish a goal that the design team didn't consider. Early involvement of general contractors is vital to integrated design efforts; they can offer early pricing of design alternatives and consult on the constructability of new approaches.

In green building, contractors are specifically tasked with pollution prevention, eliminating runoff of sediment from construction sites through such practices as silt fencing (the black plastic or cloth mesh you see at most well-run construction sites), seeding and mulching, sediment traps and basins, along with earthen dikes. LEED awards points for five major influences of construction on environmental quality:

- Reduction of site impacts from construction staging by keeping all equipment and soil disturbance within specified limits to avoid soil compaction.
- Construction waste recycling of at least 50% of materials, with extra points awarded for 75% and 95% waste diversion. This not only keeps materials out of landfills but recovers valuable products for recycling. In most urban areas, contractors are discovering they can recycle or recover more than 90% of construction waste and that this is economically beneficial for them, given the high costs of landfilling. Recycling such items as cardboard, metal, brick, acoustic ceiling tile, concrete, plastic, clean wood, glass, gypsum wallboard (sheet rock), carpet and insulation is surprisingly simple. On tight construction sites, in some cities, wastes can be co-mingled in a single dumpster and sorted offsite at a local recycling center.
- Indoor air-quality maintenance during construction through best practices such as keeping ductwork, carpets and other absorptive surfaces covered and out of the weather and dust-free.
- Indoor air-quality assurance before occupancy by conducting a two-week building flush-out with 100% outside air and changing all filters before occupancy, or by conducting a test of key indoor air-quality contaminants to make sure they are below threshold levels for health effects.
- Monitoring the activities of subcontractors to make sure that specified low-VOC paints and coatings, adhesives and sealants are actually used on the project without substitution.

Contractors document their best practices as part of the LEED certification process. Without their active and often creative cooperation, LEED projects would not be successful. Turner Construction, the largest commercial builder in the US, expects to achieve a goal of 50% waste diversion in 100% of their projects by the end of 2007, despite the difficulties of working in many cities without active recycling programs or financial incentives such as high landfill fees.[26]

Cool Roofs

Approximately $40 billion is spent annually in the United States to cool buildings, one-sixth of all energy consumed annually. Black and dark-colored roofing materials can dramatically increase a building's cooling load. Energy-efficient roofing systems, also called cool roofs, can reduce roof temperature by as much as 100°F during the summer, and thereby reduce the building's energy requirements for air conditioning. Cool roofs reflect the sun's radiant energy before it penetrates the interior of the building.[27] In fact, keeping the building roof cool helps reduce the urban heat-island effect, in which cities are markedly (4°F to 8°F) warmer than the surrounding countryside, leading to greater use of energy for summertime air conditioning.

Cool roofs provide a number of potential immediate and long-term benefits to building owners, including lower utility bills for air condition-

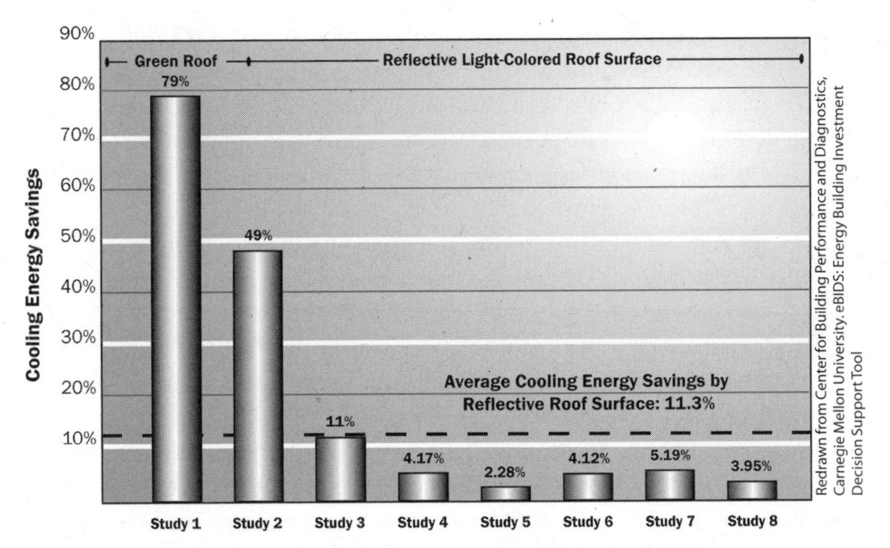

Cooling energy savings by cool roofs. Savings range from 4% to about 80% in a variety of studies and environments.

ing, smaller air conditioning systems, lower roof maintenance costs and longer roof life. Cool roofs help to meet energy-efficiency goals in building codes and help address a community's heat-island effects. Cool roofs are certified by the Cool Roof Rating Council (CRRC).[28] Designers, builders, consultants and owners are showing increasing awareness of how the radiative properties of roofs contribute to buildings' thermal performance. The CRRC recognizes only roofing-product radiative-property tests performed by independent laboratories.

The LEED rating system gives one credit point to cool roofs that cover at least 75% of a roof surface and have a Solar Reflectance Index (SRI) of at least 78 for a low-sloped roof and 29 for a steep-sloped roof. The SRI measures the roof surface's ability to reflect solar heat. A standard white roof (80% reflective) has an SRI of 100, while a standard black roof has an SRI of 0. These criteria closely mirror that for an ENERGY STAR-rated roof. ENERGY STAR-qualified reflective roof products can reduce peak cooling demand by 10% to 15% and can reduce building energy use by up to 50%.[29]

Costs of Green Buildings

As we showed earlier, a main barrier to implementing green buildings has been the perceived cost increases for green measures. It is true that many of the earlier green projects in the 2000 to 2005 period were more costly. This is largely because the transition to new methods of design and construction involves a lot of social learning that is accompanied by construction mistakes, poor designs, unproven new products and a myriad of reasons leading to extra costs. By 2005 and especially in 2006, however, many design and construction teams had done enough green projects to start lowering costs to more conventional levels.

In 2006 the developer of a large LEED Platinum project in Portland — a very complex, 412,000-square-foot, mixed-use medical facility — reported a cost premium (net of local, state and federal incentives) of about 1% on a $145 million project.[30] Now, this developer had designed and built 30 prior LEED projects and used a very experienced architect and engineering team, already well-versed in green building methods. But their success does point to the fact that future green buildings can be built without any initial cost premium, once design and construction teams garner enough experience.

What determines the cost of a green building?

- First and foremost, it depends on what the design team and owner are trying to achieve. If it's a LEED Platinum building, they most likely will use green roofs and photovoltaics, two expensive additions to a project that may not be included in a LEED Silver or possibly even a LEED Gold project.
- Second, it depends how early in the process the project decides to pursue sustainable design and construction. As we show in the section on integrated design, it's best if that decision is made as early as possible, even during the site selection process, so that a building can be properly oriented, with a rectangular shape that allows for good daylighting and efficient passive solar design measures.
- Third, it depends still on the experience of the design and construction team with green buildings; the more experience, the less the cost premium based on both fear of the unknown and lack of knowledge about sourcing green products, for example. Less-experienced teams often use green building consultants to help them out with their first project, to accelerate the learning curve.

Integrated design often leads to creative solutions that allow teams to "tunnel through the cost barrier" and design a more energy-efficient building at a lower initial cost.[31] Typically, this is done by having the architecture do some of the work of cutting energy use, as well as heating and cooling a building with daylighting, shading devices, highly efficient windows, orientation and heavy mass construction. Green buildings can also cut other project costs by saving on infrastructure investments and connection charges for storm drainage and sewage connections through total water system management. Often, by thinking strategically in the first 30 days of a project, you can influence 65% of total costs by assessing a broader range of options, making choices among key cost drivers and having a clear vision of results. This puts a premium on thinking (vs. doing), a concept that many Americans may find challenging.

One of the most widely cited studies of the costs of green buildings was done by the international cost-consulting firm Davis Langdon in 2004 and updated early in 2007. Using their own proprietary database of actual building costs, and comparing 45 LEED projects with 93 other non-LEED projects, Davis Langdon discovered that green building costs (for three types of common projects — libraries, academic classrooms and laboratories) were statistically no different than conventional building costs when normalized for year of completion (taking cost inflation out of the analysis) and location (reflecting the variation of building costs by locality).

Their work showed that the major cost driver is the building program, that is, what the building is designed to achieve. A simple branch library in the suburbs might be fairly cheap to construct, but a downtown main library will likely be much more costly, on a dollars-per-square-foot basis. You can find a large big-city downtown library by a name architect that costs $500 per square foot, as well as one that serves the same function and costs only $230 per square foot.

The figure below shows the results of the most recent Davis Langdon study for ambulatory care facilities (one of five categories with enough data from which to draw firm conclusions).[32] The 2007 update included additional project types and more cost data, all standardized to Sacramento, California, mid-2006 costs. The conclusions of the study were unchanged: certified green buildings don't cost any more than conventional buildings, on a per-square-foot basis. What matters most: the building's design objectives.

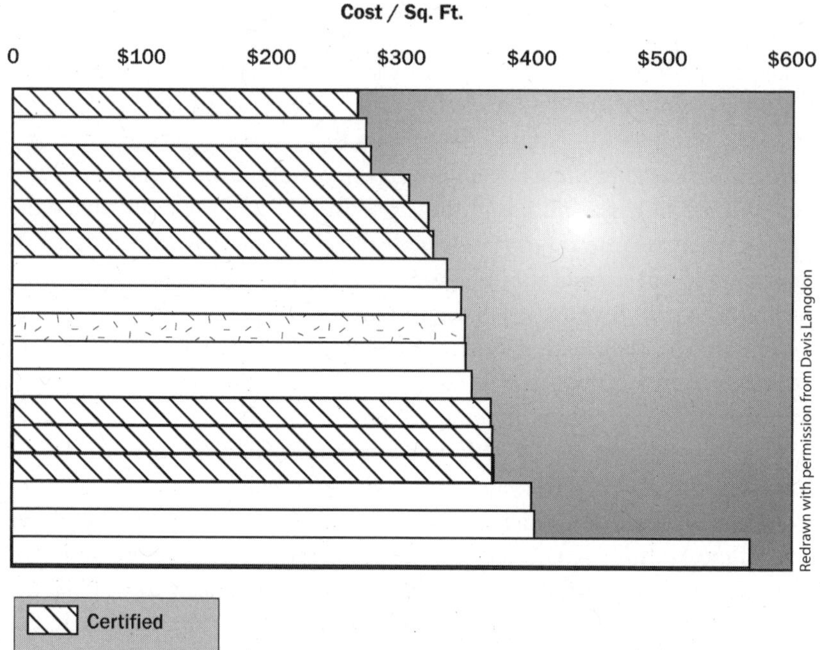

Cost of ambulatory care facilities., green vs. non-green.

Cradle-to-cradle Design

"Cradle-to-cradle" design was introduced in 2002 by architect William McDonough and chemist Dr. Michael Braungart as a method for evaluating products that could be safely used without any harm to people or the environment, based on known data. The evaluation criteria for products include material properties, specifically toxicity and carcinogenicity, persistence and toxicity in the environment, and use of heavy metals; material reuse potential, either in recycling or composting; efficient or renewable energy use, including use of 100% solar income in manufacturing; water use, stormwater and wastewater discharge in manufacturing; and instituting strategies for social responsibility as evidenced by third-party assessments and certifications.[33]

Based on a basic ecological concept that "waste is food" and the sound environmental precept that we should not be creating toxic materials that eventually wind up in the environment and in our bodies, the cradle-to-cradle idea seeks to get away from the idea that we can live in a throwaway society forever, creating thousands of new single-purpose chemicals with unknown health and environmental effects.

One of the early successes of this effort was creating a fabric for a chair manufacturer that was durable and attractive but that could be composted at the end of its useful life. Another product developed from this point of view is a commercial carpet system designed not only for sustainability in manufacturing and use, but also for recycling all of its components.[34] Another large carpet and flooring manufacturer, Interface Inc., a multi-billion dollar (revenues) company, has been on a journey to sustainability since 1995 and has committed to have zero impact on the environment by 2020.[35]

Leadership in regulating chemicals to reduce environmental and human impacts is generally found today in the European Union (EU), representing nearly 500 million citizens. At the beginning of 2007, a new regulation titled REACH (registration, evaluation and authorization of chemicals) was adopted, requiring registration and selective evaluation of more than 30,000 existing chemical substances, as well as new ones. In terms of electronic wastes such as computers, two directives adopted in 2003 require manufacturers to dispose of consumers' used electronic equipment free of charge and prohibit the export of hazardous waste to developing countries for disposal.[36]

C

In 2006 two scientists argued:

> In the 1970s and '80s, the United States effectively set many global product standards for consumer and environmental protection. Today, Europe is playing this role, while US government and industry oppose the resulting standards in Europe and in international arenas. Critics of the European Union's policies estimate costs in billions of dollars, while defenders argue that any increased costs incurred by manufacturers have previously been borne by consumers, the environment and waste contractors handling thousands of toxic substances.[37]

Daylighting

Daylighting is an aspect of green building design that should be ubiquitous; without adequate daylighting, people will not perform well and will not be healthy. For building plans, this implies a design that is no more than 66 feet wide, front to back, or about 33 feet to a window from any workstation. This is a standard design requirement in many places in Europe, where people's health is placed before economic efficiency. Looked at another way, a building should be oriented so that the long axis is east-west; this allows for maximum daylighting, from both south- and north-facing windows.

Daylighting's benefits are immediately apparent; people see better and feel better whenever there is natural light for reading and working. Good daylighting design can employ skylights, north-facing windows on the roof, a central atrium, light shelves to bounce light into a space while shading windows from the summer sun, and other techniques. Good daylight-

Daylighting at University of Oregon, Lillis Business Complex, Eugene, Oregon, designed by SRG Partnership.

ing is always indirect, without glare. Daylighting is usually combined with electric lighting, so that there is a constant lighting level, typically 30 foot-candles at the desktop, or there is task lighting provided for each work-station.

According to a report from Carnegie Mellon University analyzing daylighting research, "Eleven case studies have shown that innovative day-lighting systems can pay for themselves in less than one year due to energy and productivity benefits...the ROI [return on investment] for daylighting is over 185%."[38]

A California study of the impact of daylighting examined 73 stores of a chain retailer, of which 24 had daylighting. The results:

> The value of the energy savings from daylighting is far overshadowed by the value of the predicted increase in sales due to day-lighting. The profit from increased sales associated with daylight is worth at least 19 times the energy savings."[39]

● ● ●

Density

Recently I visited the largest green home development in the country, Civano, in Tucson, Arizona. Started in the late 1990s, it consists now of some 600 homes; developers expect to eventually construct 1,500 homes. The Civano development operates by a strict energy and water conservation code that cuts annual bills for electricity (think hot, dry, desert-climate air conditioning) by 50% and annual water use by 60%, compared with similar suburban developments in the area.[40] Civano is also denser than most suburban developments. It has a community feel to it, resulting from two public swimming pools, a community center, several small businesses (including a wonderful nursery) and narrow streets that keep down the heat in summer and serve as traffic-calming devices. Density as a sustainable design virtue reduces energy use from automobiles and allows for a more walkable community.

Certain environmental benefits of density are unquestionable. Some have even argued that New York, especially Manhattan, is the greenest city in the country (certainly for per capita use of energy and gasoline) because so many people take public transit, not only to work, but for everyday errands, shopping and leisure activities. My late father-in-law lived his entire adult life in New York City. Well into his 90s, after a car accident forced him to give up driving, he got along just fine by walking, taking the bus or subway, occasionally hailing a cab for city travel or hiring a limousine for day trips. Lots of New Yorkers don't own a car at all because of all the costs and trouble of ownership.

Having lots of high-rise residential and commercial buildings is also energy efficient, because the ratio of exterior walls to total occupied space is lower, reducing the potential for gaining or losing energy. On the basis of annual energy-use per square foot of floor space, such buildings can be much more energy efficient than a single-family home or a low-rise office or apartment. They can also support a qualified operations and maintenance staff, helping to keep equipment in good shape.

The drawback to cities is, of course, that they are largely devoid of nature. Imagine a New York City without Central Park, or any large city without parks and street trees. Additionally, cities tend to have lots of sun-absorbing asphalt, brick and concrete surfaces, which heat up during the day and don't cool off rapidly at night. This means air conditioning is required on summer afternoons and evenings, even when air temperatures should be dropping and breezes increasing. Many cities are now 8°F to 10°F hotter on summer evenings than they were 30 or 40 years ago, causing urban energy use to rise.

But cities may also be healthier than suburbs because people tend to walk more. One of the classic comedic movie scenes I recall is in Steve Martin's *L.A. Story*. It opens with him pulling his car out of the driveway one morning and driving to the house next door, getting out and going in for a cup of coffee. Having grown up in Los Angeles, with its dominant "auto-erotic" car culture, I can tell you this imagined scene isn't too far from the truth! We are all much less healthy these days because of excessive driving and infrequent walking.

● ● ●

Design

Design is the mantra of our times. In architecture, celebrity designers such as Frank Gehry, Renzo Piano and Santiago Calatrava are international icons. Management guru Tom Peters began talking about the "design decade" several years ago; he says, "Design is the seat of the soul."[41] In his view, every business needs to incorporate the essence of design thinking:

elegance with economy. With noted architect Michael Graves designing for Target and Martha Stewart for K-Mart, and stores such as Design Within Reach popping up everywhere, economical yet elegant design is in. Green design works in the same way. Getting high performance against a green rating system such as LEED doesn't have to be ugly or break the bank. One definition of superior design is "goodness of fit": how well does the final product function to meet its requirements?

Another definition from the 1970s that still works for good building design is "long life, low energy, loose fit." In his book, *How Buildings Learn: What Happens After They Are Built*, citing many historical examples, Stewart Brand argues eloquently that most buildings will go through many different uses (housing, office, retail, restaurant, etc.) during their lifetime.[42] Therefore, designers should recognize that flexibility and durability are prime design virtues for green buildings. In Oslo in 2002, during an international green building conference, I found that many Norwegian buildings, especially low-rise, were built with no internal load-bearing walls, meaning that it was easy to reconfigure an office building into apartments and vice versa, because one didn't have to worry about the structural effects of moving walls around.

Modern office buildings, built essentially as "see-through" buildings with only a core and shell, when outfitted with daylighting design, underfloor air systems and simple movable wall partitions, also meet this criteria of loose fit. They can be reconfigured in a few days for new users and, over time, could become apartments, condos or hotels on various floors. The reverse is happening to old hotels in good business locations that are transformed into offices for start-up firms. What's needed now is for green building design to embrace other virtues such as long life and low energy.

Good design also implies an "economy of means," or using fewer resources to do the job by employing more elegant strategies. In this respect, passive solar design, natural ventilation, daylighting and other green design measures epitomize good design. Using natural energies such as sun, wind and rain, before importing resources from hundreds or thousands of miles away, is also the essence of good design.

The concept of "Design for Environment" (DfE) is beginning to permeate the design community. In DfE, companies examine the long-term environmental effects of sourcing, processing, distribution and eventually recycling their products. Looking forward to the day when all manufacturers will be forced to take back their products for recycling, DfE incorporates "design for disassembly" and recycling of all components into new products.

Displacement Ventilation

Displacement ventilation is the name given to a number of techniques for letting natural forces distribute air in spaces by having cooler air displace warmer air in a space, much as filling a bathtub with cold water will eventually displace hot water flowing out the drain. Displacement ventilation has a number of virtues for the open-plan office environment, or cubicle farm. It uses far less energy than fan-forced ventilation (the ubiquitous overhead air diffusers that put a cold draft right on top of you), it moves contaminants out of spaces because air is not recirculated in spaces, and it allows people to adjust their own temperature and airflows inside of cubicles from low-velocity air diffusers located in floor tiles.

Displacement ventilation is one name given to underfloor air distribution (UFAD) systems that are beginning to see widespread use in the US. An update of the older computer-room floor, UFAD systems also accommodate electrical wiring and data cabling, enabling workstations to be easily moved as companies and agencies redistribute people in an office. The difference is that the moveable floor, a concrete tile about two-feet

Tate Access Floors, Inc.

This nine-panel mock-up shows the components of an access flooring system: the access floor with modular wiring, both a variable-air-volume system (VAV) and a swirl diffuser for underfloor air, and one-to-one-fit modular carpet tile.

square weighing 40 pounds, is typically 14 inches to 16 inches above the base floor, versus the older-style flooring at 6 inches, to allow for easier airflow under the floor.

Displacement ventilation works on the principle that more-dense cooler air falls and less-dense warmer air rises, something we all know but seldom appreciate. Typically, displacement air is introduced into a space at 62°F or 63°F, 7°F or 8°F warmer than the traditional 55°F of overhead diffusers, which saves energy in summer because the outside air doesn't need as much cooling to be usable. In some climates such as the maritime Pacific Northwest, for most of the year, air is cooler than 62°F during the day, so there is no need for mechanical air conditioning to get cool air into a space. Engineers use what's called an "economizer cycle" to bring in cool outside air whenever possible. Since most office buildings require cooling year-round, owing to the density of people and all the electrical equipment used, this is an effective strategy. As the warmer air rises, it exits at return-air grilles typically located high on a wall. In effect, with a 9-foot or 10-foot ceiling, displacement ventilation allows cooling only of the "occupied zone," i.e., up to about 6.5 feet, saving energy on cooling the entire volume of a room, as a traditional overhead ventilation system would do.

Eco-efficiency and Eco-effectiveness

Architect and sustainable design expert William McDonough exhorts green building designers not to be content with just "doing less-bad" designs that put off the day of reckoning for excessive energy and water use, but to design buildings and cities that are "positively good." When our focus is primarily on "eco-efficiency," that is, reducing our negative impacts, we are not likely to achieve design breakthroughs. For example, saving 20% of the energy of a standard building is a virtue, but energy use still creates lots of carbon dioxide emissions and pollution from electric power production. If we save 30% of the water use of a standard building, we are still using far more water than the building receives as rainfall.

Some experts tell us that our environmental impacts have to be reduced 90% or more to begin to reverse the decline in the Earth's supportive ecosystems and to relieve the stress on energy and water resources. Such a "Factor 10" building is a long way from our current focus on "Factor 1.5" buildings that might reduce impacts of building construction and operations by 33% on average. (At this time, a good LEED-certified building reduces water use by about 30% and energy use by 30% to 50%, compared with the average of all buildings.)

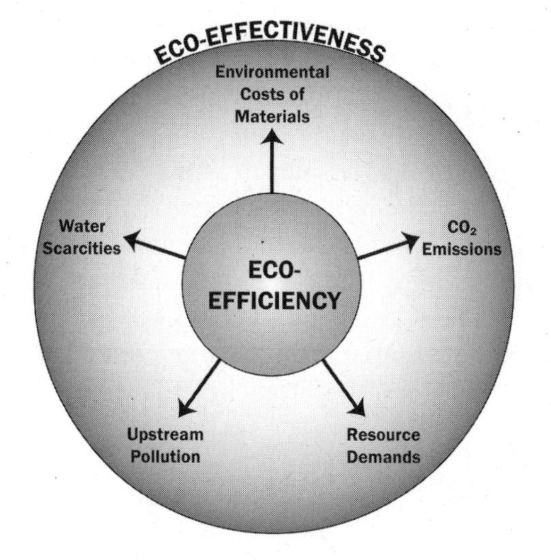

Eco-effectiveness considerations force us to include all the various environmental and social effects of our design, construction, operations and development decisions.

Economists have long analyzed the "externalities" of modern life, wherein a factory, for example, is more profitable when it is able to unload its pollution and resource depletion on the environment, without having to pay for all the consequences. One can think of the past 30 years of pollution control regulations as an attempt to make business and government "internalize" the full external costs of their pollution, so that they would decide not to create it in the first place.

In an "eco-effective" analysis, one would "internalize the externalities," for example, by performing a life-cycle analysis of all materials produced, including their upstream (cost of materials, cost of transportation, type of labor) and downstream (recyclability, reusability) environmental and social costs. The figure above shows how this approach might look conceptually. Sustainability implies that, together, we have to agree to live primarily on "natural capital" (renewable resources and biodiversity) for a long period of time, using non-renewable resources at a much slower rate, perhaps eventually not at all.

Ecological Footprint

The concept of the "ecological footprint" was introduced in the early 1990s by a Canadian researcher, William Rees, and popularized by his Swiss colleague, Mathis Wackernagel.[43] In an elegant conceptual leap, they looked at all human activity and asked the simple question: How many Planet Earths would it take to support human activity at current levels of consumption, pollution and resource depletion? The answer for 2003 was 1.25 Earths. That created an immediate conceptual problem, because we only have one Earth to work with for the foreseeable future. The ecological footprint of the average American would occupy the resources of about five to ten Earths, the average Canadian about four to eight Earths.[44]

According to the Global Footprint Network:

> Humanity's *Ecological Footprint* is over 23% larger than what the planet can regenerate. In other words, it now takes more than one year and two months for the Earth to regenerate what we use in a single year. We maintain this *overshoot* by liquidating the planet's ecological resources.[45]

Using the calculator supplied by the Global Footprint Network, you can evaluate your own consumption patterns to determine how your lifestyle stacks up against sustainability criteria.

Green buildings are one way to reduce the ecological footprint of human activity, particularly at higher levels of LEED certification. Living buildings that produce all of their own energy onsite, recycle all of their waste products, subsist entirely on natural rainfall and last hundreds of years begin to mimic natural systems.

Another approach to the ecological footprint comes from a system called The Natural Step, originally developed in Sweden in the 1980s and now in use in 11 countries.[46] Buildings that use The Natural Step assessment framework go beyond LEED to consider the full range of impacts from product manufacture, transportation, building construction and continuing operations. A modified version of this framework might ask such questions as:

- Can the Earth replace what I take? (Use no products faster than nature can produce them, either by using recycled-content materials or relying on renewable energy and natural rainfall.)
- Am I poisoning the Earth, water or air (Are my activities causing pollution either upstream or downstream of the building activity?)
- Do I respect the biodiversity of flora and fauna (Does this project result, directly or indirectly, in ecosystem degradation?)
- Are the choices I make fair and equitable? (Do the benefits of green buildings accrue to all occupants?)[47]

These standards are demanding and not likely to be met by most of today's green buildings, but they point us toward the development of restorative buildings and offer a clear direction for future green building design.

● ● ●

Education

"We shape our buildings; thereafter they shape us," attributed to Sir Winston Churchill, aptly describes one of the more important aspects of green buildings — what they teach the occupants and the public about how buildings should be designed, built, renovated and operated. Properly done, green buildings reflect designers' and builders' expectations about high-performance design and can teach the public what to ask for in their own homes and workplaces. The people who live and work in them can easily take on the values of the buildings they inhabit.

Public education is seen as a crucial component of the green building revolution, so much so that LEED gives an innovation point for projects that include significant public education. Your next green building project can foster an understanding of sustainable design through such measures as:

- Educational signage throughout the building, even in the restrooms, describing the various measures taken for energy conservation, water conservation and recycling.
- User handbooks and training sessions that inform building occupants, especially for buildings with atypical systems.
- Websites, intranets and other electronic means that educate users throughout the design and construction process and also serve as a repository for information about building systems.
- Public tours not only educate and inspire general audiences about green buildings but can serve to reinforce for occupants the special measures taken to ensure their well-being.
- Using see-through panels can take some of the mystery out of building systems, by exposing them to public view.
- Visible monitors that track energy use, photovoltaic power production and water use can give occupants satisfaction they are doing something positive for the environment.

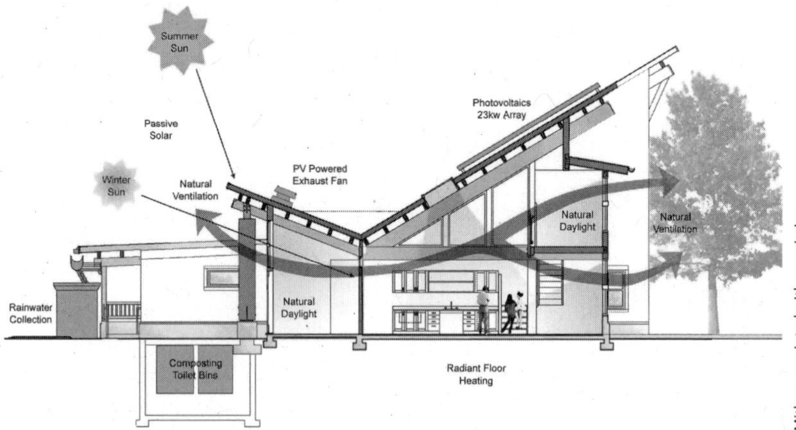

Mithun architects designed the Islandwood Environmental Education Center on Bainbridge Island, Washington, a LEED Gold-certified project, to teach sustainability with its building systems.

e

Energy Conservation

This is a vast topic, so here we're just going to introduce a few key concepts. The basic notion is that energy conservation is the cheapest power we can buy, or to use Amory Lovins' term, it generates "negawatts," negative demand on the power grid and fuel supply. A 2007 report from the American Solar Energy Society concluded that energy efficiency (conservation) and renewables could contribute an average of about 1,200 million metric tons per year of carbon reductions by 2030, putting the US on a path to achieve the country's goals under the Kyoto Treaty. Energy efficiency could contribute 57% of the total, renewables 43%.[48] For the building sector alone, the potential reduction is 200 million metric tons per year, about one-sixth of the total reductions needed in carbon dioxide emissions. That's how important building energy conservation technologies, designs and techniques are to our future.

What are some of the most important energy conservation measures that should be designed into our commercial and residential buildings? Here's a list of the more cost-effective measures.

- Better insulation of buildings and homes, including floor, window and ceiling insulation.
- Plugging the leaks, especially in homes, so that heated or cooled air doesn't escape before it's used.
- Better glazing, including double-pane, "low-e" coated glass for all buildings.
- More efficient air-conditioning systems, with higher SEER (Seasonal Energy Efficiency Ratio) ratings, along with radiant cooling systems.
- Waste-heat recovery systems from exhaust air.
- Natural ventilation and operable windows to allow nature to provide heating and cooling.
- Improved lighting technologies and greater use of daylighting and LEDs.
- Higher-efficiency boilers and radiant heating systems, including tankless water heaters.
- Better methods to control moisture in buildings, to allow comfort with less cooling energy.
- Reducing losses of conditioned air with better duct sealing techniques.
- Ground-coupled heat pumps (geothermal heating systems) that reduce whole-building energy consumption by 30%.

- Rightsizing (downsizing) of HVAC systems so that they operate more efficiently.
- High-efficiency condensing boilers for water heating and tankless water heaters that supply hot water at point of use, only on demand, reducing losses through gas flues.
- Distributed energy systems that allow for far greater total energy-use efficiencies.
- Changes in building codes to require more stringent minimum energy-efficiency levels.
- Reducing the required size of building ventilation fans through a variety of techniques.
- Premium-efficiency motors with variable-speed drives to better match input energy with loads.
- Water conservation measures to reduce energy use for water heating.
- Carbon dioxide sensors to control ventilation levels based on occupancy avoid wasting energy.
- Occupancy sensors to turn off lights and HVAC systems when spaces are not in use.

Engineering (a Sustainable World)

Building design focuses so much on architecture that many people forget that it's the engineers who make all the building systems work, from structural systems to heating, ventilating and air conditioning (HVAC) to plumbing, lighting, electricity, technology systems, external water supply and waste disposal, fire protection and life-safety systems. About 15% to 20% of the construction cost of a typical commercial building is taken up by engineered systems, not by the building structure or furnishings, while an additional 15% to 20% comes from the civil construction outside the building.[49] It's easy to see that the engineer's role is critical to achieving high-performance building design with economy.

Engineering is all about economy of resources. Someone once told me that a good engineer is someone who can do for $1 what a mediocre engineer can do for $2. What would a more substantial green building design role for the engineers look like, and how would that lead to economy of means to achieve high-performance results? We discuss this in several sections, including Rightsizing, Integrated Design and Energy Conservation,

Sustainable engineering vs. "post-modern" engineering

Category	"Post-modern"/ Conventional Engineering	Sustainable/High-performance Engineering
Buildings	Suburban greenfields; New buildings preferred	Urban infill; Adaptive reuse of building stock
Energy use	Meet energy code; Reduce energy use vs. code; Make incremental improvements; Buy power from local utility	Exceed code by 50%; Reduce absolute energy use; Develop new systems and methods; Use on site power such as co-generation
Economics	First cost is major driver; Look only at project economics	Life-cycle cost analysis; Triple-bottom-line thinking
Ventilation	Forced ventilation; Sealed windows; High-pressure central systems	Natural ventilation; Operable windows; Low-pressure distributed ventilation
Climate control	Design with components; Narrow temperature band; Consider only HVAC system economics	Design whole systems; Expanded temperature band; Look at health and productivity of the workforce
Water use	Specify efficient fixtures	Reclaim and reuse rainwater/ graywater
Stormwater	Convey off site to rivers and ocean	Detain, retain, recharge, re-use on site
Wastewater	Convey off site to treatment plant	Treat and reclaim for on site use
Materials selection	Environmental effects not considered	Life-cycle assessment of materials; Use recycled materials such as fly ash for concrete

but consider the above table that illustrates how engineering needs to change to promote sustainable design. I contrast sustainable or high performance engineering against the conventional state of engineering design for "post-modern" architecture.

Feng Shui

Feng Shui is an ancient Chinese system for the harmonious placement of buildings on a site and for the placement of rooms and objects within a building or office. The goal is to allow the free movement of a subtle energy called "chi" that permeates the world. (Most martial arts systems are based philosophically on the movement of this subtle energy.) In Feng Shui, cities and homes are seen as intimately linked with both the landscape and the cosmos. According to one expert, in China, "Every activity involved in making the city, living in it and participating in its life reminded a person of the forces that acted upon their world. They became aligned with those forces and gained nourishment from them."[50]

Currently, I doubt that many US architects know about Feng Shui, but it should become a major part of green building design in this country, particularly at the residential level, over the next decade. In China today, hardly any building is built or office designed without consulting with a Feng Shui expert.

The Chinese system employs the five elements of wind, water, fire, metal and wood to harmonize surroundings with the energies of the Earth. Grounded in traditional Chinese cosmology and religion, it's based in a country (with a geography like the United States) where the cold winter winds bear down from the north and the warm spring and summer breezes blow in from the south; therefore, it could certainly be used in this country. We find elements of Feng Shui here in the popularity of table fountains for home and office and in the use of fish tanks at the entrance to most Asian restaurants.

In some ways, we already use these concepts without knowing it. The need for daylight and fresh air, natural ventilation, views to the outdoors, places to see long distances (prospect) and places for retreat (refuge), the use of plants and water features in buildings, fountains and koi-filled ponds — all these speak of implicit knowledge and use of Feng Shui design precepts.

In my own home in Portland, I saw first-hand the benefits of using Feng Shui guidance. We had an older home with a large wooden support beam running across the living room and bedroom ceiling. This gave a very heavy feeling to the rooms, as if there were a weight on your head. So we hung a feather from the beam in both rooms, and almost immediately the oppressive feeling was relieved. At my home in Tucson, where a cul-de-sac sometimes results in vehicles facing directly at the house, I installed a mirror by the front door to deflect the harmful "chi" force emanating from them. Feng Shui also instructs us where to place water features in the house and how to avoid designs that let positive energy escape, such as a front door that has a direct line of sight to a back door. Feng Shui instructs the design of an entire house, including the placement of each room, and also informs site selection and lot layout.

Most architects in the US are rationalists and materialists and don't yet know how to incorporate a Feng Shui approach into their buildings. I contend, however, that we are just beginning to learn how buildings affect people on a deeper psychological and spiritual level. If we are serious about building healthy buildings (for healthy people), in which we can live, learn, play and work productively, architects and builders need to learn more about Feng Shui design principles.

Furniture and Finishes

Furniture plays a significant role in green offices and homes. The life-cycle impacts of the materials used in chairs, tables, desks, partitions and similar systems, the source of the wood products and the ecological footprint of fabrics are all considerations in green interior design.

One of the easiest ways in which companies can start on the journey to sustainability is to evaluate their furniture and furnishings purchases and to incorporate such criteria. The LEED for Existing Buildings system explicitly incorporates "environmentally preferable purchasing" policies into the rating system. LEED for New Construction also rewards furniture made from salvaged and reclaimed materials; recycled-content materials, rapidly renewable materials, certified wood products and composite materials that are free of urea-formaldehyde resins. Consider the new Steelcase Think chair, which is up to 99% recyclable by weight. Disassembly for recycling takes about five minutes using common hand tools. The chair has up to 44% recycled content. It holds the "NF Environnement" label in France for environmental quality and is Greenguard Indoor Air Quality certified in the US.

Formaldehyde-free Materials

Urea-formaldehyde (UF) is a suspected human carcinogen.[51] When it's present in the air at levels of at least 0.1 parts per million (ppm), acute

f

health effects can occur including watery eyes; burning sensations in the eyes, nose and throat; nausea; coughing; chest tightness; wheezing and skin rashes.[51] It is also a ubiquitous resin used in most composite wood and agrifiber products, to hold the pieces of wood or particle board together. Trouble is, UF stinks, causes irritations and is almost impossible to get rid of, even after airing it out for a considerable period. If you try to buy any conventional furniture that's not made of solid wood, you'll experience the UF smell. Recently, I made a trip to a very large, well-known international furniture retailer in the Phoenix area and was quite distressed to find almost no wood furniture that didn't smell of UF.

There are substitutes, including phenol-formaldehyde, that emit far less formaldehyde gas. So why don't manufacturers and designers use them? It's almost always related to cost and performance of the resin binder. In a LEED building, there is a credit for the use of composite wood and agricultural fiberboard products, including glulam beams, that don't use UF resins. In commercial buildings, where designers are specifying products that will influence the air quality for others, shouldn't they use furniture and other wood products that are odor-free, without toxic fumes?

Steelcase, Inc.

The Think chair by Steelcase is the first product to ever receive Cradle-to-Cradle™ Product Certification from McDonough Braungart Design Chemistry (MBDC).

g

Global Warming

As part of climate change caused by human activity, global warming seems to be well underway. Average temperatures have risen (0.5°F) over the past 100 years and seem destined to rise 1°F to 2°F or more by 2050, without dramatic changes in current patterns of energy use and continued global population and economic growth.

A number of gases contribute to global warming, including carbon dioxide from fossil fuel burning (from homes, cars, buildings and industry) and methane from bovine flatulence (believe it or not, this is major contributor). However, other gases also have global warming potential, including chemicals in common use such as chlorinated hydrocarbons used in refrigerants, nitrous oxide (laughing gas), trichloroethane (a solvent) and other halogenated compounds (those containing chlorine, fluorine and bromine atoms).

The 2007 Fourth Report of the Intergovernmental Panel on Climate Change (IPCC) put the odds that human activity is causing global climate change at more than 90%. That report, along with Al Gore's movie, *An Inconvenient Truth*, and the British Government's "Stern" report, have moved public opinion in the advanced economies toward fully recognizing the threat of global warming and pressing their governments to take immediate action.

Global warming could cause the Arctic ice cap to melt, releasing huge amounts of fresh water into the North Atlantic, the rapid splitting off of parts of the Antarctic ice sheet and melting of glaciers worldwide. The latter could greatly increase droughts because there is less winter rain held as snow and, therefore, reduced river flows during the summer. Vast portions of Southeast Asia depending on the Mekong River for sustenance, in turn fed by Himalayan glaciers, could be affected. Low-lying areas such as Miami and New Orleans could become like Venice, Italy, inundated a good part of the year and basically uninhabitable. Island nations could see their habitability decrease dramatically in the next few decades. Warmer seas could increase the severity of hurricanes in the Gulf of Mexico in years ahead, even if their frequency does not increase.

Global warming is observable, and the consequences predictable and dire. The connection with fossil fuel burning is getting more provable scientifically each passing year. Green building advocates strongly favor "zero-net-energy" buildings as one way to cope with this dramatic challenge to current civilization. Globalization is not only an economic phenomenon, it's also an ecological phenomenon. We are one interconnected world, at levels we don't really fathom or can't accept. Green buildings provide a way to ensure sustainability of current economic, cultural and political institutions.

Green Globes

Green Globes is a green building rating system originally developed in Canada. Its US sponsor, the Green Building Initiative, introduced it into the US in 2004:

> Green Globes for new construction was adapted from a system that is widely used in Canada, where it is one of only two green building rating systems recognized by the Canadian federal government. Under the trade name *Go Green Comprehensive*, it is also the basis of the Building Owners and Managers Association of Canada's national energy and environmental program for existing buildings.[52]

Green Globes has found some market interest, partly in answer to complaints that LEED certification costs too much and is too complex. In response, Green Globes created a Web-based project information system. In the beginning, Green Globes was supported heavily by the forest products industry, because it gave credit to wood products certified by an alternative certification program, the Sustainable Forestry Initiative.[53] At the end of 2006, Green Globes had registered about 100 commercial projects for certification, roughly 3% of the market, and had certified about 10, approximately 2% of the number of LEED certifications. In the residential sector, by contrast, Green Globes has teamed up with the National Association of Home Builders and local home builder associations to offer its rating system to the new home market. At the beginning of 2007, 11 local home builder associations provided the Green Globes assessment system for use in their members' projects.[54]

Green Globes has its adherents, of course, who claim that the Web-based system is easier to use than LEED (which is also currently Web-based). A 2006 study by the University of Minnesota found that LEED and

Green Globes overlap in more than 80% of their credits.[55] The study also concluded that "Green Globes employs a rating criterion that reflects life-cycle thinking and covers the entire life cycle of building materials," but that neither LEED nor Green Globes "successfully addresses functional relevance with regard to materials selection."

The Green Globes rating system consists of seven categories with the following weights on a scale of 1000 points. It takes 350 points to get one Green Globe, 550 points to get two, 700 points to get three and 850 points to get four.[56]

Project Management	50 points
Site	115 points
Energy	360 points
Water	100 points
Resources and Materials	100 points
Emissions/other impacts	75 points
Indoor Environment	200 points

Green Guide for Healthcare

The Green Guide for Healthcare (GGHC) is a best-practices guide for healthy and sustainable building design, construction and operations for the healthcare industry. It is based on the LEED rating system (and licensed from USGBC) and has been strongly supported by architects designing healthcare projects as well as by the industry. Unlike LEED for New Construction, GGHC is divided into separate sections for construction and ongoing operations. The version 2.1 pilot program has more than 115 healthcare facilities participating, as of early 2007, representing more than 30 million square feet of new construction projects. GGHC version 2.2 was released early in 2007. In the fall of 2007, LEED for HealthCare will be released, to complement GGHC version 2.2. The GGHC protocol has 42 core credits (vs. 32 in LEED) and 96 total core points (vs. 64 in LEED). The green operations section of GGHC has also been used as a stand-alone best-practices guide for healthcare facilities.

The business case for greening healthcare facilities is best summed up with a few items:

- A $100,000 annual savings in operating costs for energy and water is equivalent to the profit from generating $2.7 million in new revenue because hospitals are constantly facing pressures on increasing revenues, costs savings from green buildings make fiscal sense.[57]

- Greening healthcare facilities is completely in line with the core values of health and healing. This commitment to building and operating healthy facilities fulfills the physician's first dictum: "First, do no harm." Greener facilities resonate with staff values at all levels.
- Greening healthcare facilities provides better outcomes for patients. Getting them home faster is beneficial for them and economically advantageous for the hospital or clinic.
- Greener facilities also help with recruitment and retention of key staff, such as nurses, by providing green relief from a hospital's stressful environment.
- Greener hospitals are great for public relations, showing the community that the institution cares for more than its immediate surroundings. This is important, since almost 90% of hospitals are nonprofits and compete fiercely for public support in many arenas.

In terms of the cost of implementing green building criteria in hospitals, architect and cost management specialist Lisa Matthiessen says:

> The cost situation is going to be about the building management process. You have to be committed to green building, and you have to share that vision with all the stakeholders. You have to be clear about its goals and expectations...and you have to write an integrative design process plan.[58]

Interior designer Jan Stensland with Kaiser Permanente, one of the oldest and largest healthcare organizations in the country, says, "There is a learning curve to all this, but the bottom line is the people factor. We are in this business to serve people. The best, most heartwarming result of these changes is the staff's appreciation of our efforts."[59]

● ● ●

Green Home Building Guidelines

Most green home programs follow the "Green Home Building Guidelines" of the National Association of Homebuilders (NAHB).[60] Developed earlier in this decade, the NAHB's green guidelines form the basis for dozens of local certification programs run by local home builder associations. (There are also established green home certification programs run by local utilities, such as Austin Energy in Texas, the oldest program in the country.) The NAHB's guidelines divided the green home assessment program into seven guiding principles, using a point-based rating system.

- Lot design, preparation and development.
- Resource and materials efficiency.
- Energy efficiency.
- Water efficiency.
- Indoor environmental quality.
- Operation, maintenance and homeowner education.
- Global impact of products used, including low-VOC paints and sealants.

For certification at the Bronze, Silver and Gold levels, the NAHB guidelines require minimum point totals in each of the seven areas of concern, plus 100 additional points from sections of a builder's choice. As expected, minimum energy-efficiency points are by far the largest single component of the system.

Civano Community

Civano community in Tucson, Arizona, is an early (1990s) example of a New Urbanist walkable, green home development with clear energy-efficiency and water-conservation guidelines, narrow streets and local businesses near the town center. Individual homes were constructed and sold by four local builders. A second 600-home phase is underway in 2007, using the same guidelines.

Green Power

Green buildings (as well as you and I) have the option of buying some or all of their electricity from a green power provider. In this context, green power is renewable energy produced somewhere else and then transmitted to the project site (figuratively). There are dozens of such suppliers in the US, most of which are wholesalers of wind power, some solar power and some biomass, geothermal or low-impact hydro facilities. My own local utility, Tucson Electric Power, sells two packages of GreenWatts power, one 99 % wind and 1% solar, the other 90% wind and 10% solar (at a 20% premium). Tucson Electric Power operates one of the largest photovoltaic (solar electric) power plants in the country, a 5.1-megawatt array capable of producing 7.5 million kilowatt-hours per year of electricity, enough to power 500 Arizona homes.[61] I purchase 2,400 kilowatt-hours per year of this power, about 20% of my annual use, for an annual premium of $186, or about 7.75 cents per kilowatt-hour. For commercial users, typical renewable energy premiums are lower, about 2 cents per kilowatt-hour.

In LEED projects, one energy-efficiency point is allowed for purchasing two years' worth of green power from a recognized supplier, provided that the purchase offsets at least 35% of the annual electricity use. The

© iStockphoto.com/Dean Bergmann

Most of the green power in the US comes from wind power.

Center for Resource Solutions' (CRS) Green-e program is used to certify renewable energy sources for the purposes of the LEED credit.[62] In 2005 the CRS verified that green power production grew 43% over 2004, to a total of 5.2 million megawatt-hours (5,200 million kilowatt-hours) of renewable energy generation. According to the Center, "Green-e certified renewable electricity products were sold in 48 out of 50 US states, and 215,000 MWh (about 4%) was purchased by companies who contracted to use the Green-e logo to promote their commitment to certified renewable energy."[63] In addition to Green-e certified providers, other sources of green power such as Green Tags,[64] renewable energy certificates (RECs) and tradable renewable certificates (TRCs) are eligible for LEED project credits if they satisfy the Green-e program technical requirements.

Green Products

Green products and green building materials have some or all of the following characteristics:

- Low embodied energy (total energy to make, distribute and use the product).
- Recycled-content and/or can be easily recycled after use.
- Uses renewable resources from forests and agriculture.
- Harvested, extracted and processed within 300 to 500 miles, supporting regional economies and reducing transportation.
- Energy efficient in operation, such as high-R-value glazing products.
- Low or no environmental impact in manufacturing as well as in use and disposal.
- Durable (this is an often overlooked attribute of green products, but quite important).
- Minimize waste in manufacture and use, including engineered wood products from scrap.
- Positive social impact, contributing to health and well-being.
- Affordable, preferably lower cost than conventional alternatives with fewer green characteristics.[65]

What makes a product green? That's a question that admits no clear answer, particularly as green products become more mainstream (i.e., when you can buy them at Home Depot, Lowe's or Ace Hardware, rather than from specialty distributors or retailers). Some examples of green products are:

- Innovative lighting controls and new uses for ultra-high-efficiency LEDs (light-emitting diodes).
- Renewable energy systems, or processes promoting the use of renewable energy.
- New forms of building glazing that save energy by reducing heat gain/loss.
- Smart irrigation controls that save water by only applying it when the soil is dry.
- Low-flow showerheads, low-flow sinks, water-free urinals or dual-flush toilets.
- Permeable paving products that allow water infiltration onsite.
- Modular green roof technologies that allow lower costs and ease of installation.
- Engineered wood products for decking, sheathing and framing lumber that use far fewer materials than conventional products, or use materials that would formerly have been considered waste .
- New products using certified wood in furnishings, engineered wood and furniture.
- Polished concrete floors that reduce the need for finished floors in a building.

The LEED rating system is certainly driving the development and production of green products, even though it doesn't reward specific products with any individual points in the rating system. Rather, in many cases, it is the cumulative total of all products with a certain green feature that garners the particular point. In response to growing market demand, many new green products are introduced each year.[66]

Green Roofs

Green roofs are one of the most obvious and visible commitments that a green building can make. In addition to providing habitat for plants and animals, a green roof can assist with stormwater management and can provide some additional buffering of the environment, to reduce heating demand in winter and cooling demand in summer. Green roofs are found everywhere in North America, from cold, wet northern places like Toronto and Chicago to hot, dry southern places like Phoenix. The City of Chicago has more than 200 green roof projects underway, including one on the City Hall.

Green roofs come in two varieties, intensive and extensive. An intensive green roof is thicker and can support a wider variety of plants; for example, a LEED Gold apartment project, the 27-story Solaire in Battery Park City, is home to a rose garden on the roof of the 19th floor. However, intensive green roofs add more weight and require more irrigation and maintenance, so most of the projects use extensive treatments, in which the soil layer is thinner (less than four inches) and typically composed of lightweight materials such as perlite.

Green roofs can also be designed as a second-floor amenity over a ground-floor retail podium in office buildings or residential high-rises in the cities. Available to office workers or residents, the green roof provides a park-like space in the midst of a city.

Green roofs are not cheap. Typical costs range from $10 to $20 per square foot. For a 15,000-square-foot green roof, that would be $150,000 to $200,000. However, for a large project with multiple stories, a green roof might represent a significant amenity at a cost of 1% or less of the total project cost. In a project with high-level LEED goals, a green roof can help

William McDonough + Partners

The Ford Motor Company River Rouge Plant green roof provides bird habitat as well as stormwater management benefits.

with open-space goals, thermal comfort, stormwater management and reducing the urban heat-island effect, cutting air-conditioning costs in summer and holding water from small to medium-sized storms for later release.

Green roofs are used in many applications, including commercial, industrial, government and residential buildings. In Europe they are widely used for their stormwater management and energy savings, as well as their aesthetic benefits. Green roof systems may be modular, with drainage layers, filter cloth, growing media and plants already prepared in movable, interlocking grids, or each component of the system may be installed separately in layers.

A good example is the Ford Motor Company's River Rouge plant. Lying at the center of the revitalization project, this new assembly plant represents Ford's efforts to rethink the ecological footprint of a large manufacturing facility. The design emphasizes a safe and healthy workplace with an approach that reduces the impact of the plant on the external environment.

The keystone of the site's stormwater management system is the plant's 10-acre (454,000-square-foot) living roof, the largest in the world. This green roof is expected to retain half the annual rainfall that falls on its surface. It will also provide habitat, decrease the building's energy costs and protect the roof membrane from thermal shock and UV degradation, extending its life. Very quickly, local birds discovered a safe place to lay eggs!

In describing the project's strong economic justification, architect William McDonough discovered that Ford was prepared to spend $48 million on a conventional stormwater management system to handle the runoff from the roof and parking lots of the facility. His design cost about $13 million ($29 per square foot of roof), including the green roof and a parking lot with gravel filters and bioswales (planted drainage ditches), followed by constructed wetlands treating stormwater runoff, saving Ford $35 million along the way. He said, "It takes three days for water to flow from the plant to the river, and it's purified naturally along the way.... It took [Ford's] board about a minute and a half to approve the project."[67]

High-performance Buildings

Many people have begun using the term "high-performance buildings" instead of "green buildings" or "sustainable buildings" because they want to emphasize what is gained from these projects, not what is given up.[68] High-performance also appeals to Americans; we want everything turbocharged and super-sized. A high-performance building is one in which energy and water efficiencies are high, indoor air quality is high, recycling rates are high, etc. This is a much easier concept to explain to most executives than a green building, which still sounds vaguely like a tree-hugger term. In my view, high-performance buildings are those that save at least 50% of the energy use of a standard building, compared with a database called the Commercial Buildings Energy Consumption Database, last updated in 2003.[69] (The next survey will be conducted in 2007.) The table below shows an example of the information that's available from this database.

From this table, you can see that the energy-use intensity of buildings varies by primary activity; it also varies by date constructed (though not as much as you might think), geographic location (heating/cooling climate), height (stories), operating hours and public/private ownership type.

Average Energy Consumption of Buildings

Principal Building Activity	Energy Consumption (thousands of BTU per sq.ft. per year)
Office	93
Public Assembly	94
Education	83
Health Care (inpatient)	249
Health Care (outpatient)	95
Food Sales (grocery store)	200
Food Service	258
Lodging	100
Retail (non-mall)	74

Commercial Buildings Energy Consumption Survey, Energy Information Administration

There are so many factors entering into building energy use that engineers have to employ elaborate computer models just to compare a high-performance building with a standard or typical building. Next time you plan to have a building constructed or renovated, insist that designers deliver a high-performance building!

● ● ●

Historic Preservation

Historic preservation and green buildings go together nicely. If green buildings are all about sustainability, what could exemplify this value better that reusing an older building, making it suitable for another 50 to 100 years of active use?

In Portland, the Jean Vollum Natural Capital Center is a great example of building renovation. Under the leadership of Ecotrust, a regional non-profit, the 100-year-old two-story warehouse was transformed into a modern office building. A partial third story was added onto the original build-

Jean Vollum Natural Capital Center, second LEED-NC Gold building in the US, Portland, Oregon, designed by Holst Architects.

ing, and a green roof covers 50% of the roof area. An outdoor deck was added to host local receptions. The first floor contains a retail store, a local coffee house, a pharmacy and a pizzeria. Partly as a result of this project, the surrounding neighborhoods began to sprout high-density residential, retail and restaurant uses. In recognition of its improved energy efficiency, use of recycled content, focus on locally sourced materials and extensive use of certified wood products, the Center became the second LEED Gold-certified project in the US when it was completed in 2001. As a center for environmental education, Ecotrust's conference room is the most frequently booked meeting place of its type in the entire city, hosting hundreds of events each year. Everyone who visits the center can see the value and beauty of building renovation first-hand.

Buildings on the National Register of Historic Places are often good candidates for green building renovations. Such buildings may qualify for 20% federal tax credits, but they come with significant restrictions on renovation activity; while any building more than 70 years old (e.g., built before 1937) can qualify for a tax credit of 10% of expenditures, without such significant restrictions. The first renovated building on the National Register to receive a LEED-NC Platinum rating was the Gerding Theater, also in Portland, which involved the renovation of an 1891 armory building into a performing arts theater.

● ● ●

Homes, Green

What's a green home? If you're in the market for a new home, you should be asking for a green home, one that, at a minimum, saves at least 30% of the energy of a conventional new home, has water-conserving fixtures and uses non-toxic finishes. Increasingly, regional and national homebuilders are offering such projects. A recently completed Northern California project by Lennar, one of the ten major US homebuilders that deliver about 35% of all new single-family homes, shows this potential. The project in Roseville will integrate photovoltaic systems and upgraded energy-efficiency measures into 450 new homes slated to be built over the next two years in cooperation with Roseville Electric, a city-owned utility. The PowerLight SunTile® solar electric system,[70] a roof-integrated technology, will be installed as a standard feature on each home, resulting in significant utility bill reductions for the homeowners, and the generation of

Photo by Dan Morrison, courtesy of *Fine Homebuilding Magazine*

Cannon Beach home, NAHB Green Custom Home of the Year, 2005, is also a zero-net-energy home, built in accordance with advanced ecological design principles. Design by Nathan Good Architect.

completely emissions-free electricity. In addition, by reaching certain energy-efficiency goals, the Lennar-built homes may qualify (through 2008) for a $2,000 federal tax credit.[71]

Another project by a well-established regional builder is Carsten Crossings in Rocklin, California (near Sacramento), built by The Grupe Company. This project was LEED-certified by the LEED for Homes pilot program. The homes are more energy efficient than the exacting California state requirements and also include onsite solar electricity that can reduce electrical bills by up to 70%. Standard energy-saving features also include tankless water heaters, "low-e" windows, enhanced attic insulation and foam-wrapped building envelopes.[72]

What makes a home green?

- It should save at least 30% of the heating and cooling energy of a conventional home through additional insulation, better glazing, better construction methods, exhaust-air heat recovery (in cold climates) and more-efficient HVAC systems.
- It should be built according to an established rating system such as the NAHB model Green Home Building Guidelines (or their local home-builder association or utility equivalent), ENERGY STAR or LEED for Homes.
- It should be certified by an independent third party using the chosen rating system.
- It should supply some of its energy from renewable sources, either solar hot water or photovoltaics.
- It should save water with water-conserving fixtures such as dual-flush toilets.
- All appliances should be ENERGY STAR rated.

There are custom homes that take the concept of a green home even further, by having a green roof, 100% of energy supplied by photovoltaics and solar thermal systems; using only FSC-certified wood products; using recycled-content, salvaged and locally sourced materials throughout; in essence trying to get as high up on the LEED for Homes rating scale as possible. Such homes include homeowner education to maximize the effectiveness of the energy conservation and water conservation systems. Shown at left is a custom home in Cannon Beach, Oregon, that received the NAHB Custom Green Home of the Year award in 2005. Its design also incorporated The Natural Step principles (see section on the Ecological Footprint), which was very important to the homeowners.

FXFOWLE ARCHITECTS, PC

The Helena apartments, New York City, designed by FXFOWLE ARCHITECTS, PC

Elsewhere, we have described projects by developers that include mid-rise and high-rise condominiums and apartments. For any building above three stories, the LEED for New Construction (LEED-NC) rating system can be used to assess a green residence. Several dozen such residential projects have already been certified under the LEED-NC rating system, including a Portland apartment building that received a LEED Gold certification in 2007. The nicest feature of this apartment, in which my wife and I were among the first tenants, was the healthy indoor air quality, including the fact that this was a non-smoking building.

● ● ●

Hybrid Vehicles

Hybrid vehicles and green buildings might seem an odd match — one goes down the road, the other stays in one place forever — but a focus on reduced energy consumption unites them. Hybrid car sales in the US in 2006 reached 200,000, about 1% of the overall vehicle sales. However, an early 2007 survey of auto industry executives revealed that 71% believe hybrids will become more important as a US market force, with between 200,000 and 500,000 cars selling in 2007.

However, it's clear that hybrid sales are also tied to fuel prices. For example, between August 2006 and December 2006, gasoline prices declined 24% and hybrid sales declined 31%.[73] Federal tax credits for hybrids also sweeten the financial benefit of reduced gas use. In 2007 the tax credit for hybrids ranges from $650 to $3,150, depending on the make and model and the total number of hybrids sold since the credit began. The credit phases out in 2009.[74]

Many companies in the green building industry that want to distinguish themselves in terms of their commitment to sustainability are embracing hybrids for their own fleets and sometimes even subsidizing employees to purchase hybrids for their own use. Certainly, among certain "eco-elites," having a Toyota Prius has become a status symbol, a visible sign that the owner cares about lowering gasoline consumption and is willing to spend a little more to do so. But, buying and operating hybrid and highly fuel-efficient cars and trucks is more than a status symbol or a way to get a LEED point in a new project; it is also a component of a more sustainable future for any business. Why not become more competitive by lowering your cost of operations through reduced fuel use?

In the LEED for New Construction rating system, a credit point is granted under Alternative Transportation for providing low-emitting and fuel-efficient vehicles for 3% of the full-time equivalent (FTE) occupants and providing preferred parking for those vehicles. If a building has 200 full-time occupants, for example, an owner could qualify for this credit by providing six of these vehicles for use, perhaps for carpools. (To qualify, cars have to be Zero Emission Vehicles or score 40 or more on a green score. For 2006, Honda and Toyota hybrids qualified.)[75]

With more than 60% of America's oil coming from foreign sources, ranging from the stable Saudi Arabia to the very unstable Iraq and Iran, to the politically questionable sources in Latin America, why shouldn't we vote for greater national security through reduced fuel purchases? Cutting gasoline consumption is one of the few things you can do on a daily basis that has such a dramatic and immediate impact on national security. Don't forget we live in a world where peak oil production is likely to come not only in your lifetime, but within your working life. Worldwide, we continue to extract far more oil each year than we discover, meaning that we are drawing down the available resource faster than it's being replenished. One industry expert claims that the last time we found as much new oil as we pumped was 1988, nearly 20 years ago. The imbalance widens each year. In 2005, according to this analyst, we only replaced 20% of the oil that was pumped with proven new supplies.[76]

Incentives

Given a choice, most people prefer the carrot to the stick; at this point in the development of green building methods, techniques and technologies, incentive systems seem a better approach than mandates. Incentives allow the private sector to experiment with a vast array of methods for achieving various levels of energy-efficiency and LEED certification. By combining all of a building's environmental attributes into a point system, LEED makes it easy to trade off among various components of a building while still achieving a specified result such as Silver, Gold or Platinum.

However, green building advocates and local and state government leaders are not going to wait around for the private sector to construct high-performance buildings. By 2010, if not sooner, we are going to see incentives coupled with mandates, as green buildings and green homes move into the mainstream. The issues of combating climate change are too urgent and too political to wait a generation for the private sector to start constructing and operating buildings in a sustainable manner. But for now, incentives are the preferred method for accelerating the growth of green buildings.

Examples of State and Local Green Buildings and Renewable Energy Incentives

Type of Incentive	Some Examples
Income tax credit for green buildings	Oregon and New York
Income tax credit for solar energy systems	Arizona, New York, New Mexico, Utah and Oregon
Property tax abatement for LEED Silver or higher ratings	Nevada, Michigan, Minnesota and Massachusetts
Sales tax relief for solar units	Arizona, Florida, Georgia, Idaho, Iowa, Maryland and Ohio
Priority processing of building permits	Chicago and Los Angeles
Density bonuses for LEED projects	Seattle and Arlington, VA
Utility subsidies for photovoltaic systems	California, Arizona, Alabama, Colorado, Texas, Hawaii and Florida
Grants and loans for renewable energy systems	Alabama, Connecticut, Massachusetts, Minnesota, New York, Ohio and Tennessee

Directory of State Incentives for Renewable Energy

State and local governments offer many incentives for green buildings and renewable energy systems. As you can see, there are a lot of public policy options and incentive opportunities for promoting green buildings at any level of government.

Integrated Design

Integrated design is the main method used by green builders to design high-performance buildings on conventional budgets. How is this possible? If I make a piece of equipment more efficient, isn't it going to cost more? Yes, it is. But if I make it more efficient *and* design the building to use less energy overall, then that equipment is going to be both more efficient *and* smaller, thereby saving cost.

Here's a practical example. Most office buildings don't use shading devices external to the building; instead they rely on tinted glass to reduce glare. But the sun still enters the building, leading to excess heating in summer, requiring more cooling. Most office building air-conditioning

Traditional "Design, Bid, Build, Hand Over" Process

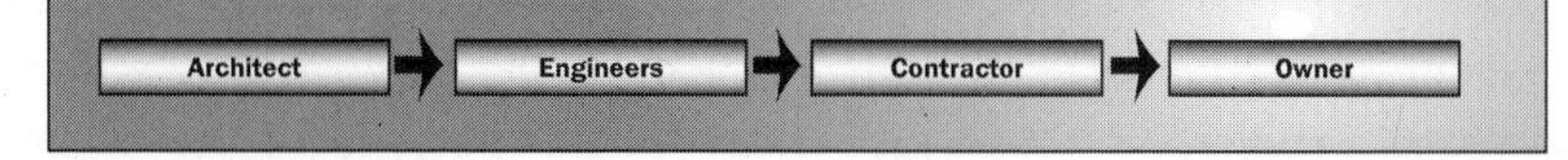

Integrated Design

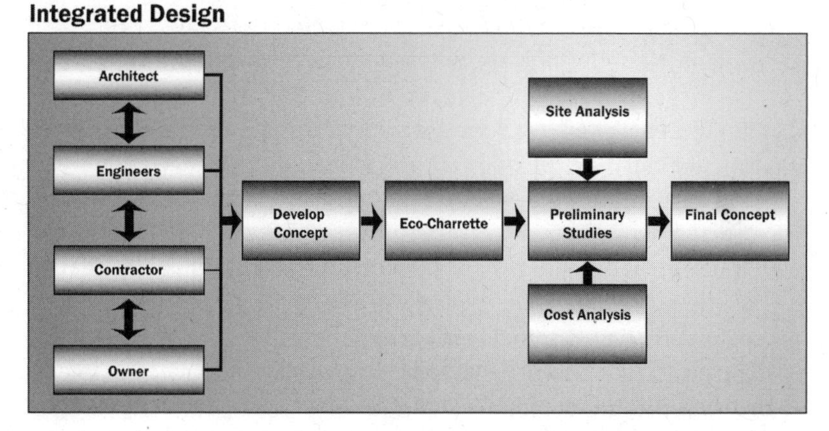

Integrated design needs to be front-loaded in the design process to identify and capture opportunities.

systems are sized for the absolute peak summer afternoon cooling demand. What if you could reduce that demand by shading the south-facing and west-facing (and in some southern climates, east-facing) glass, so that the sun never enters the building during the summer, at least from 10 am to 4 pm? That would require a lot less cooling. But it would cost more for the external shading devices (plus mess up the clean post-modern façade of the building). But would it cost more than you would save by downsizing the air-conditioning system? In general, the answer is no.

So the basic principle of integrated design is that we have to look at the whole building's energy and water use, not just focus on individual systems in isolation from each other. Easier said than done, however, without a strong commitment to an integrated design *process* by the building owner, architect, engineers and other key members of the design and construction team.

The basic elements of integrated design include the following activities, all knitted together by excellent facilitation and communications skills:

- Select the Team: Team members need to know not only a lot about their own field of expertise, but also how their decisions might affect all of the building's systems and components. Typically, the owner's representative or project manager and the key contractor team members are part of the design effort, much more so than in conventional building.
- Set the Goals: It's a truism in business (and life) that teams can perform at very high levels if they are asked to. It's also true that "what gets measured, gets managed," so that measuring performance (such as achieving a LEED rating of a certain level) is a critical factor in driving design teams to innovative, cost-conscious decisions.
- Assign Champions: It's also a truism in business that someone has to "own" the results. If an architect tells a mechanical engineer to cut building energy use from the norm by 50%, that engineer has to agree to make this happen, without overspending her budget. High-performance buildings have a champion for each area of design, construction and operations.
- Optimize the Design: No one project is just like another; each design has unique requirements, typically driven by the purpose of the building, the owner's or developer's expectations, budgets and schedules. Therefore, each design needs to be optimized for the specific circumstances. This often requires multiple iterations of the design, tradeoffs between various building elements, best done in the early stages before much of the detailed design cost has been incurred.

- Follow Through: The best players in sports never take their eye off the ball, and the best building projects have a very strong follow-through, to make sure that all of the great ideas from the early stages of design get incorporated into the project. This can be difficult, because often the building's conceptual designer hands off the detailed design to another architect or engineer, or is occupied with several other projects happening simultaneously.
- Commission the Project: As discussed previously, commissioning makes sure that the design intent is actually translated into the building systems all the way through construction and startup, so that all energy-using systems work the way they were designed. Incredible as it may seem, no one except perhaps the building owner or developer is really in charge of a building's final energy performance. Commissioning helps to bridge the gap between good intentions and actual performance.
- Maintain and Monitor: No good project goes without maintenance for long. Training of operating personnel in the building's systems (especially if they are unconventional) is critical, as is monitoring the performance of the building and finding out why performance isn't meeting expectations.

Interior Design

Green interior design is a vast topic; designing healthy workplaces, using green materials, daylighting and good lighting controls, low-toxic finishes, low-emitting carpet, cabinetry and furniture from sustainably harvested wood, lower-water-using fixtures — the list goes on. The LEED for Commercial Interiors (LEED-CI) standard quantifies these measures in a point system. By early 2007, 34 projects had been certified under the LEED-CI version 2.0, including 4 Platinum-rated projects, and 59 under the LEED-CI version 1.0 (pilot) program, including 1 Platinum-rated project. Here we highlight the key elements of three of the Platinum projects and one Gold project.

InterfaceFLOR's showroom in Atlanta has shown its commitment to sustainability with the first Platinum-certified LEED-CI project. Opened in 2004 in an existing building, the project diverted 92% of construction waste from landfills, installed low-flow fixtures in restrooms, installed

The InterfaceFLOR showroom in Atlanta, designed by TVS Architects, was one of the first LEED-CI Platinum projects.

energy-efficient lighting and lighting controls, reused 30% of furniture and fixtures, minimized VOCs used during construction and installed low-VOC carpets and furniture.[77]

In 2006 the architectural firm Cook+Fox relocated its offices in New York City and sought to redefine the firm and convey its vision for the future with a LEED-CI Platinum-rated interior remodel. Located in the 12,000-square-foot penthouse of a former department store in downtown Manhattan, the project has extensive daylighting from its 14-foot ceilings. A zoned lighting system uses energy-efficient metal-halide and fluorescent lighting, as well as high-quality ambient light with desktop task lighting. The existing air-handler, relatively new, was upgraded with a variable-frequency drive and additional filtration. A 3,600-square-foot green roof uses biophilia design principles to fight the urban heat-island effect and to provide a visual amenity for the office. Water consumption has been lowered by 40% through fixture choices. Green materials were used throughout, and the firm instituted an extensive green cleaning program. The entire staff's home energy usage, as well as the building's, has been offset through the purchase of green power credits.[78]

The Society for Neuroscience's new headquarters in Washington, DC, received a LEED-CI Gold rating in 2006, covering the improvements on

Credit: Society for Neuroscience

Envision Design's remodel for The Society for Neuroscience space incorporated a number of key elements of green interior design.

three floors of an 11-story, 84,000-square-foot building. The project reduced water use by 22%, and HVAC systems comply with increased energy-efficiency requirements. Daylight-responsive controls are used at perimeter windows. All electricity used in the space comes from green power sources (wind). Recycled materials make up 22% of the total materials cost, including 39% of the carpet, 40% of the resilient flooring, 50% of upholstered wall panels, 33% of gypsum board, 71% of ceiling tiles and 50% of office furniture. All wood comes from FSC-certified forests. Low-VOC materials were used throughout.[79]

In February of 2007 the US Green Building Council moved into its new headquarters, a 22,000-square-foot remodeled space that received a LEED-CI Platinum certification. Some of the features include: rapidly renewable bamboo flooring, reused granite countertops and non-toxic paint on the walls. Employees in the open workspace areas enjoy abundant natural daylight and operable windows, and 93% of the interior has a view of the outdoors. Office furniture from the previous space was reused, and the ceiling tiles are recyclable. All lighting products are engineered for high efficiency and lower energy use; individual task lights allow employees to control light within their work area. The new office uses 40% less water via low-flow plumbing fixtures, dual-flush toilets and water-free urinals.

i

Investing in Green Buildings

In 2006 real estate investment in green buildings grew dramatically with the announcement of a joint venture by the California Public Employees Retirement System (CalPERS) to fund future commercial real estate projects by the Hines development organization. A large national real estate developer, Hines is well-known for its commitment to ENERGY STAR and LEED certification for its buildings.[80] The Hines CalPERS Green Development Fund (HCG) was capitalized with over $120 million of committed equity and, with leverage, will have the ability to invest up to $500 million. HCG will concentrate on developing high-performance, sustainable office buildings certifiable through the LEED for Core and Shell Program.

A number of publicly traded Real Estate Investment Trusts (REITs) have also begun to make strong commitments to green buildings, including Liberty Property Trust (Philadelphia, PA)[81] and Corporate Office Properties Trust (Columbia, MD).[82] In addition, major private investors have begun to develop large green projects. Two notable examples are Gerding Edlen Development in Portland, Oregon, and Vulcan Real Estate in Seattle, Washington. By the end of 2006, Gerding Edlen registered more than 30 development projects on the West Coast for certification under the LEED system and developed the world's largest LEED Platinum building (to date, at 16 stories and 412,000 square feet), the Center for Health and Healing at Oregon Health & Science University in Portland.[83] Vulcan is developing a number of properties in Seattle's South Lake Union district to LEED standards.[84]

Investing in green buildings has begun to attract considerable attention as a form of socially responsible investing, a practice that is growing faster than overall investing. According to one socially responsible property investing expert, Professor Gary Pivo at the University of Arizona:

> We have yet to see the first real estate investment fund squarely committed to green real estate. But until such funds are created, there are some other options worth considering. One is to acquire shares in companies that commonly own ENERGY STAR-labeled buildings or have been recognized by ENERGY STAR for their conservation efforts. Examples include Arden Realty, Equity Office Properties, Hines, Brandywine Realty, Carr America, Glenborough Realty, Parkway Properties, Prentiss Properties and USAA Realty.[85]

Justice, Social

One of the more heartening events of the past two years has been the response of the design community to the aftermath of Hurricane Katrina, showing how equity considerations are a vital part of the triple bottom line for green buildings. At the November 2005 Greenbuild show in Atlanta, held annually by the US Green Building Council, many of the best minds in the world of green architecture dedicated themselves for three days to holding an eco-charrette on how to respond to Katrina's dispossessed population with affordable green housing. The *New Orleans Principles* to guide sustainable reconstruction included the following elements: respect the rights of all citizens of New Orleans; restore natural protections of the greater New Orleans region; implement an inclusive planning process; value diversity in New Orleans; protect the city from future flooding; embrace smart redevelopment; and provide for "passive survivability" (defined as the ability to survive without outside services such as electricity or water supplies for some period of time) in the event of a future crisis or systems breakdown.[86]

Concurrently, the Congress for the New Urbanism held a design charrette for New Orleans homes and came up with some manufactured-housing solutions that would allow houses to be built for about $35,000, with enough room for a family, storage for their goods and a front porch to greet the neighbors.[87]

Sustainable design proponents and green building designers are working to make sure that the benefits of green buildings reach the entire population; one way to do this is to build green affordable and public housing. Many architects are working on such projects. For example, in September 2004 the Natural Resources Defense Council (NRDC) and Enterprise Community Partners (formerly the Enterprise Foundation) launched the Green Communities Initiative, a five-year, $550 million commitment to build more than 8,500 environmentally friendly affordable homes across the country.[88]

Pugh + Scarpa Architects

The LEED Gold-certified Colorado Court located in Santa Monica, CA, designed by Pugh + Scarpa Architects.

Certified as a LEED Gold project, Colorado Court, a 30,000-square-foot, 44-unit single-resident occupancy (SRO) apartment building in Santa Monica, California, was designed to be 100% energy independent, using passive solar design strategies, an onsite gas-fired cogeneration system and solar electric power. The project will save each unit $150 per year on utility bills and is strikingly beautiful as well, showcasing the aesthetic potential of polycrystalline silicon PV panels.

Knowledge

It's often commented that modern society is "smart, but not wise," that we know how to do lots of things right, but we don't necessarily do the right things. Green buildings develop from a knowledge base not only of how to make energy-efficient buildings, but how to do so in the least harmful and most resource-efficient way. In their highly influential book, *Natural Capitalism*, authors Paul Hawken, Amory Lovins and L. Hunter Lovins point out how we've learned to "optimize" each subsystem of a building, but in a way that "pessimizes" the performance of the overall system. In other words, engineers can make a building relatively comfortable with air conditioning and forced ventilation, no matter how poorly the architect has designed the building. That's smart, but is it wise?

My late father-in-law, a mechanical engineer for five decades, lived to be 100; he liked to say that an engineer is "an economist with technical training." In other words, the engineer's goal should be to minimize resource use to accomplish a specific economic purpose, e.g., make buildings healthy, comfortable, productive places to live, work, play and learn. Most engineers have this knowledge or some part of it, but the time-pressured system of building development, design and construction conspires to make most of them into glorified equipment specifiers, not whole-systems building engineers.

Wouldn't it be better to start with the idea of a building that heated and cooled itself because of its orientation to the sun, its use of shading and overhangs, its understanding and use of radiant heating and cooling from the mass of the building itself, its use of natural ventilation solutions to let air move through the building with minimal fan energy, etc. and then (and only then) add the smallest mechanical and electrical systems needed to take care of the remaining demands?

This line of thinking represents the essence of the sustainable design revolution: the use of both traditional "bioclimatic" building designs along with the best of modern systems, controls, materials and technology to develop a new "indigenous" building style for each region. Why shouldn't a building in Tucson resemble a saguaro cactus, exposing the minimal amount of surface area to the sun and the water-draining dry warm air, and one in Portland look more like a tall, majestic Douglas fir, comfortable in its cool, cloudy, wet surroundings?

Arizona buildings ought to be heavy, massive affairs, with radiant-cooled, shaded surfaces and with no direct sun penetration except in winter. In summer the same surfaces would absorb heat from the air during the day and re-radiate it at night to the atmosphere, when no one would be around. They should harvest all of their energy from the abundant sun and conserve whatever water that falls.

Pacific Northwest buildings on the west slope of the Cascades should admit maximum daylight from above, turn a closed north face to the cold winter breezes but admit the low-angle winter light and allow for cross ventilation during the mild spring and fall shoulder seasons. They should recycle all of their abundant rainwater and be able to harvest natural earth energies for both heating and cooling.

LEDs (Light-emitting Diodes)

LEDs are a revolutionary new lighting technology that reduce energy consumption, allow lighting to be programmed by computer and permit wide variations in lighting color. LEDs use chips, not bulbs, so they emit a lot less heat than incandescent or even fluorescent lamps. Made with computer chips, they are easily dimmable and programmable.

A good example is a new product from Herman Miller that's winning design awards. Shown below, the Leaf personal light offers maximum lighting options with less material, in an intriguing, organic form and with controls that invite human participation. The proprietary technology also addresses the most vexing problems in existing LED solutions — light intensity and heat buildup. It draws only eight to nine watts of power, about 40% less than a compact fluorescent (CFL) bulb with the same light output.

"Leaf is designed to give the user a full spectrum of choices to express light's magical and sensory variations," says designer Yves Behar. "It lets human senses become engaged by allowing the user to choose the intensity and color of light which best suits a functional need, mood or location."[89] Leaf was developed according to Herman Miller's demanding Design for the Environment (DfE) protocol, emphasizing sustainable

Herman Miller, Inc.

The Herman Miller Leaf Light.

processes, materials and recyclability. Leaf's environmental impact is perhaps most profound through its use. On average, Leaf's LEDs consume little power, last up to 100,000 hours and cut energy use significantly compared even to CFLs.

LEDs are already in major use in traffic signals, as cities and counties throughout the country are using them to replace standard bulbs. In addition to saving energy, the LEDs' long life reduces maintenance costs for replacing burned-out bulbs by almost 90%. A lighting design colleague of mine recently used LEDs for highlighting a light-rail bridge, programmed to put on a light show every time a train passed over. The possibilities for using LEDs in lighting design are endless!

LEED (Leadership in Energy and Environmental Design)

LEED is the leading green building rating system in the US for commercial, institutional, and mid-rise to high-rise residential buildings. It is developed, trademarked and owned by the US Green Building Council. At the end of 2006, nearly 5,000 buildings had registered their intention to use the LEED rating system for new buildings, renovations, tenant remodels and existing buildings, a 50% increase over 2005 year-end totals. Almost 650 million square feet of buildings were involved in LEED-registered projects, representing close to $78 billion worth of construction. By the end of 2006, according to the US Green Building Council, more than 650 projects had already been certified, an increase of 66% over the 2005 year-end totals. In an industry that grows typically by 5% per year, this increase in green building projects is astounding. The stated goal of LEED is to transform the building industry by introducing rating systems that reflect scientific knowledge, leading-edge architectural and engineering design approaches and best practices in construction and development. LEED is divided into six rating systems:

- LEED for New Construction (and major renovations).
- LEED for Commercial Interiors (remodels).
- LEED for Core and Shell (typically office buildings and other speculative projects).
- LEED for Existing Buildings (the effects of continuing building operations).

- LEED for Homes (custom homes and production homes, including low-rise apartments).
- LEED for Neighborhood Development (campuses and urban districts, new subdivisions).

The dominant rating system at this time is LEED for New Construction (LEED-NC), comprising 77% of certified projects and 79% of all project registrations. (Registering a project is like getting engaged, declaring your intention to certify when a building is completed and ready for occupancy.) In my opinion and experience, LEED-NC is a very carefully constructed rating system. It was introduced in March 2000 after a two-year beta test or pilot program and has seen widespread use and popularity.

LEED for New Construction (LEED-NC)

The basic and most frequently used LEED rating system contains 32 categories of environmental design and energy concern, with 64 core points and 5 extra credit points, for a total of 69. A basic LEED-NC certified project must score at least 26 points in the categories of site, water, energy, materials and indoor environment. There are also Silver (33 points), Gold (39 points) and Platinum (52 points) certification levels. Less than 20 LEED-NC projects had been certified Platinum by the end of 2006.

Each project requires rigorous documentation that is evaluated by independent auditors. A finished LEED-NC certified project is well within the top 10% of all buildings constructed each year, in terms of its green attributes. LEED-NC certified projects also tend to be 30% or more energy efficient than their conventional counterparts, use 30% less water and have healthier indoor air, more daylighting and views to the outdoors.

LEED for Core and Shell (LEED-CS)

The LEED-NC standard relies on evaluating a completed and fully furnished building. Many commercial buildings are built as "core-and-shell" projects, meaning that 50% or more of the project is empty (a "see-through" building) when construction is completed, and the office spaces are left for tenants to build out. A core-and-shell building typically has a lobby, an elevator core, a finished external and interior structure, major HVAC, plumbing and electrical systems, parking garage and little else. These projects are evaluated using the LEED for Core and Shell standard that makes allowances for items that developers do not finish, such as lighting, carpeting, paints and similar items. The original idea was to link the LEED-CS rating with a LEED for Commercial Interiors rating, so that the entire building would ultimately be certified similar to a LEED-NC project. In practice, this does not always work out, since even committed green developers are reluctant to impose green tenant improvement standards.

LEED for Commercial Interiors (LEED-CI)

This rating system covers the environmental design and energy issues that a tenant improvement project can address, including lighting energy use and quality, HVAC energy use and controls, access to public transportation, energy use by office equipment, office furniture, other building materials choices, cabinetry, carpets, paints, furnishings and a host of issues related to the constraints of working within an existing building. Introduced late in 2004, the LEED-CI standard has registered nearly 500 projects and certified nearly 100. There are 57 potential points in a LEED-CI project; a certified project must achieve 21 points, 27 for Silver, 32 for Gold and 42 for Platinum.

LEED for Existing Buildings (LEED-EB)

This rating system attempts to deal with the continuing environmental footprint of a building or development after it is occupied. According to

Cook+Fox Architects' LEED Platinum office, the first in New York City, turned a historic penthouse space into a healthy and creative work environment. The daylit, open studio features natural materials, filtered air and a view onto a 3600-square-foot green roof.

the US Green Building Council, LEED-EB "is a set of voluntary performance standards for the sustainable upgrades and operations of buildings. It provides sustainable guidelines for building operations, periodic upgrades of building systems, minor space use changes and building processes."[90]

Most of us would think of the energy and water use in evaluating building operations, but LEED-EB broadens this perspective to include recycling rates, chemical use in landscaping and pest management, environmentally preferable purchasing policies, green cleaning and maintenance and a number of other building management issues. The long-term effect of a building on the environment is the sum of a lot of little choices that building owners and operators make throughout its lifetime. LEED-EB is the first comprehensive system to assess these effects and to suggest how to mitigate them.

Introduced late in 2004, LEED-EB registered almost 250 projects and certified about 40 by the end of 2006. LEED-EB has 32 credit categories and 85 possible points; 32 points are required for certification, 40 for Silver; 48 for Gold; and 64 for Platinum.

LEED for Neighborhood Development (LEED-ND)

LEED-ND was unveiled early in 2007 as a pilot project (beta test) rating system for assessment of up to 240 projects. Developed in a close partnership with the Natural Resources Defense Council and the Congress for the New Urbanism, LEED-ND seeks to provide a national set of standards for neighborhood location and design based on the combined principles of smart growth, new urbanism and green building. LEED-ND will certify exemplary development projects, based on evaluating location efficiency; environmental preservation; compact, complete and connected neighborhoods; and resource efficiency.[91] We expect the finished version of LEED-ND to be rolled out in the fall of 2008, after the USGBC assesses the results of the pilot program.

LEED for Homes (LEED-H)

The pilot phase evaluation of LEED-H, 2005 to 2007, will culminate in the release of version 2.0 in 2007. The pilot phase certification system contains the five basic LEED categories (sustainable sites, water efficiency, energy efficiency, materials and resources use, and indoor environmental quality) plus two categories unique to the residential situation: location and linkages (land development practices, infrastructure, community resources and compact development); and homeowners awareness and education about green buildings. The pilot phase had 108 total points; the version 2.0 system is likely to have fewer. There is also an issue of cost and verification;

because most homebuilders will not spend more than $250 to $400 per home for certification and inspection, LEED-H has had to come up with a sampling protocol for housing tracts, unlike the other rating systems for larger projects.

● ● ●

Life-cycle Assessment (LCA)

"Paper or plastic?" Everyone wants to know. And the definitive answer is "It depends." Paper means cutting down trees and operating paper mills; plastic means pumping oil, operating refineries and chemical plants. Which is less environmentally harmful? This is the type of question addressed by the field of Life-cycle Assessment, or LCA. While it is possible to use LCA to make choices among materials for specific uses, the actual superiority of one material to another still depends on how you quantify the intrinsic, cultural and economic value of a number of environmental attributes such as:

- Water pollution.
- Air pollution.
- Global warming.
- Environmental degradation.
- Ozone depletion.
- Habitat destruction.
- Human health.

One of the leading systems in the US for evaluating environmentally preferable products (EPP) for public and private purchasing is called Building for Environmental and Economic Sustainability (BEES), put out by the National Institute of Standards and Technology (NIST). According to NIST:

> BEES measures the environmental performance of building products by using the life-cycle assessment approach specified in the ISO 14040 series of standards. All stages in the life of a product are analyzed: raw material acquisition, manufacture, transportation, installation, use and recycling and waste management. Economic performance is measured using [a] standard life-cycle cost method, which covers the costs of initial investment, replace-

ment, operation, maintenance and repair and disposal. Environmental and economic performance combine into an overall performance measure."[92]

For a manufacturer, an LCA would involve making detailed measurements of energy and water use, waste generation and other environmental impacts associated with the manufacture of the product, from the mining of the raw materials used in its production and distribution, through to its use, possible reuse or recycling and eventual disposal. LCA enables a manufacturer to quantify how much energy and raw materials are used and how much solid, liquid and gaseous waste is generated at each stage of the product's life. Using one of the comprehensive assessment methods such as BEES, a manufacturer can then determine how its products stack up against direct competitors and substitutes.

According to a 2006 report prepared for the US Environmental Protection Agency:

> Life-cycle assessment is a 'cradle to grave' approach for assessing industrial systems.... By including the impacts throughout the product life cycle, LCA provides a comprehensive view of the environmental aspects of the product or process and a more accurate picture of the true environmental trade-offs.[93]

Life-cycle Cost

Life-cycle cost (LCC) analysis is a relatively straightforward way to evaluate energy and water conservation technologies that save money long into the future, but which may cost more initially. In other industries, LCC is called Total Cost of Ownership, which takes into account not only energy savings, but also future operating and maintenance costs. If you buy a cheap car or cheap appliance, generally you expect to spend more on its service than for an expensive car or appliance, or you expect that it will not last as long. Since mechanical building systems such as air conditioning last 20 to 25 years and other systems such as glazing and insulation can last the lifetime of the building, it makes sense, certainly for public agencies, non-profits, universities, schools and corporate building owners, to look at the long-term picture.

To complete an LCC for renewable energy systems, energy conservation technologies, more-efficient lighting designs, better glazing for buildings, higher levels of insulation and similar measures, you need to know these four items:

- Initial cost of baseline systems vs. more efficient technology.
- Annual operating costs of the baseline technology vs. the more efficient technology.
- Annual savings (for example, in energy and water) in today's dollars, or in future dollars.
- Some way to express future savings in present dollars (typically called the discount rate).

When these are known or can be reliably estimated, then the calculation is simple and can be expressed in one of two ways, the Net Present Value of Savings (NPV) or the Return on Investment (ROI). If the discount rate is not known, then a method known as the Internal Rate of Return can impute it.

Some engineers still like to use the phrase "payback period," meaning the time it takes to recover the initial increase in investment cost with savings. For example, if a technology costs $300,000 and saves $100,000 per year, then the payback period is three years. This approach helps with risk management (how long are you willing to wait to get your money back?), but it doesn't count savings beyond the payback period that can add significantly to the overall investment return. For instance, typically developers do not include callback frequencies and costs within the investment horizon analysis. (Callbacks include post-occupancy complaints or maintenance problems.) Yet often a green building project will have far fewer callbacks to which developers would need to respond, and these savings can have considerable economic value.

There's one further complication: many people believe that future gas and electric utility prices will increase faster than the rate of inflation. Therefore, a calculation based on energy savings at today's utility rates will *underestimate* the value of energy conservation investments. So, it may be realistic to examine the utility price trends in your state or local area to see what you think will be the future rates, especially for peak-period power demand (typically occurring on summer afternoons). With electricity demand growing at 5% to 7% per year and supply growing only 2% per year, utility price increases for peak-period electricity use appear to be inevitable.

Public agencies are beginning to require LCC for new projects. For example, the State of Washington requires it for all new public schools.[94] You can encourage your school system to do the same, if you want to develop an approach that will result in more energy-efficiency investments.

● ● ●

Light Pollution Reduction

In the 1950s, the city of Tucson, Arizona, enacted one of the first dark-sky ordinances in the US to safeguard viewing conditions at Kitt Peak National Observatory, about 60 miles away. A small town at the time, Tucson is now a metropolitan area of more than a million people, but the culture of the dark sky still holds. Light fixtures direct their light down instead of up, have lower illumination levels in general and are designed to go off during the late night hours. I live within the city limits, but on the eastern edge of urban development. In my development of more than 100 homes, there are no street lights, just small fixtures by the mailbox in front of the house. As a result, I can see lower-magnitude stars near the horizon on a clear night. Try this in a typical city; it's often hard to see any but the brightest stars most nights. I grew up in Los Angeles, and to see the stars, we had to go camping or drive far away from the city.

Lower nocturnal light levels offer benefits other than astronomical observing. In the desert, because of the intense daytime heat, many animals are nocturnal, and dark skies provide a protective cover for their feeding and hunting activities. According to another source:

> The ecological effects of artificial night lighting are profound and increasing. Each year, over four million migrating birds are killed in collisions with lighted communications towers in the United States. Dispersing mountain lions miss crucial landscape linkages because they avoid lit areas. Increased night lighting disrupts important behaviors and physiological processes with significant ecological consequences.[95]

Reducing light trespass from a project site also makes one a good neighbor. Many studies have found that lower illumination levels can also be safer, because there is less contrast between light and dark areas, allowing one's eyes to see better into the dark. For older people, whose eyes don't adjust as fast, it can take several minutes to adjust from brightly lit parking areas to darker zones.

I always wonder why large office buildings in cities have so many floors lit up during the night when no one is working. Some simple engineering, providing multiple lighting zones for each floor, would allow the janitors and other nighttime workers to have light for their work without wasting so much energy. Even corporations who pay millions of dollars to put their names on the top of all buildings may not realize how much adverse public relations they are creating by keeping the lights on at night.

LEED provides one credit point in the category of "sustainable sites" for meeting certain established criteria for night lighting levels and light trespass from a building site. These standards have been established and maintained by the Illuminating Engineering Society of North America. As one might expect, the criteria for exterior lighting vary, ranging from more stringent (for projects located in darker rural areas and park settings) to relatively less stringent (for projects located in the brightly lit centers of major cities). Interior lighting that can be seen from outside should be reduced by 50% or more between the hours of 11 P.M. and 5 A.M., or else shielded from outside view.

Lighting Design

Lighting is one of the critical components of green building design, accounting for 23% of a typical office building's energy use.[96] Lighting design is also an important factor in raising the productivity level of office workers. Over the years, recommended illumination levels have gradually come down, from the brightly lit fluorescent workplaces of the 1970s, where 100 foot-candles was the recommended luminance, to today's standard green building design of 30 foot-candles at the work surface. In modern lighting design, especially with underfloor air systems, there can be task and ambient lighting, as well as lighting displays for special effects such as highlighting public art. Of course, many people can work well in illumination below even 30 foot-candles, provided that the light is natural daylighting.

Research at Carnegie Mellon University demonstrates the importance of lighting quality for productivity. Given that costs for people's salaries and benefits are typically 100 times a building's energy costs, even a small gain in productivity can be the equivalent of paying the energy bill several times over.

Productivity gains from implementation of high-performance lighting systems.

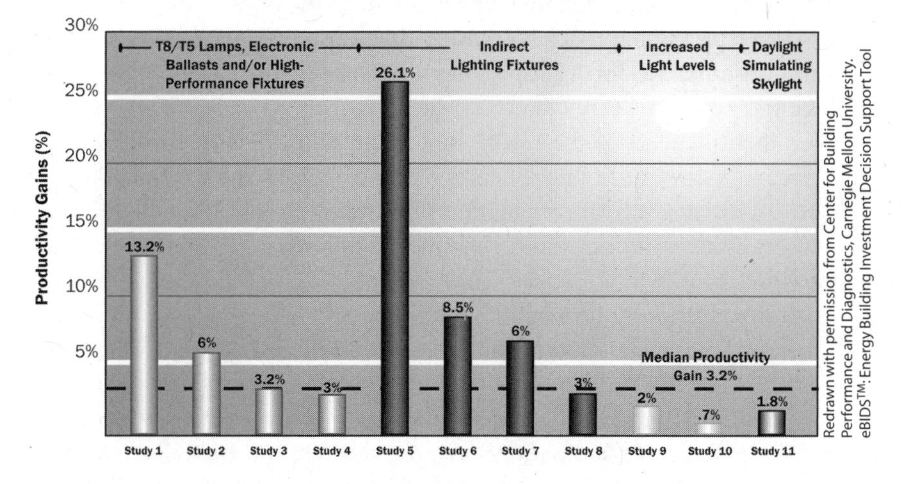

Redrawn with permission from Center for Building Performance and Diagnostics, Carnegie Mellon University. eBIDS™: Energy Building Investment Decision Support Tool

New lighting technology also has an impact on energy use and lighting quality. Elsewhere, we discuss the advent of light-emitting diodes (LEDs), a strong candidate to replace other lighting technologies in many uses. The advent of T-5 high-output lighting in the early 2000s cut 37% of the materials out of a fluorescent light by reducing the diameter from 1 inch to 5/8 inch (and provides a brightly lit lamp reminiscent of the light sabers used by Luke Skywalker and Darth Vader in *Star Wars*). On a trip to Alaska in 2007, I toured a building that effectively combines T-5 lamps with direct-indirect lighting fixtures, in which light bounced off the ceiling, as well as shone directly out the bottom of the fixtures, providing the effect of a dispersed lighting system with fewer fixtures.

Living Buildings

One of the concerns of the green building movement is how to move toward a zero-carbon footprint for buildings in which people will still want to live, play and work. Current green building rating systems such as LEED focus primarily on reducing impacts from conventional building construction and operations, but do not "guarantee" an environmentally positive outcome. To meet this challenge, leading architects have begun designing "positive impact" buildings that actually produce more energy than they consume, restore habitat and clean the water they use. In 2006, to promote this approach, the Living Building Challenge was launched by the Cascadia Region (Pacific Northwest and British Columbia) Green Building Council.

"At the heart of the Living Building Challenge is the belief that we need to move quickly to a state of balance between the natural and built environments," say the organizers.[97] According to this assessment framework, living buildings should exhibit the following performance characteristics:

- Build only on previously developed sites.
- Set aside habitat permanently, acre for acre, for each new development.
- Use net zero energy, subsisting entirely on natural flows of sunlight and wind, measured on an annual basis.
- Treat and reuse all water onsite, using natural rainfall (except for potable water requirements for drinking and washing).
- Generate almost no construction waste.
- Discharge no onsite wastewater or stormwater; treat and recycle all wastewater onsite.
- Use no materials with demonstrated toxicity to humans or the environment, including heavy metals, formaldehyde, PVC, chlorinated fluorocarbons and halogenated flame retardants.
- Use materials and services from nearby, less than 250 miles in the case of stone and aggregate.
- Offset carbon dioxide from construction activities through the purchase of carbon offsets.
- Use only wood products certified by the Forest Stewardship Council.
- Provide operable windows for fresh air and daylight in each occupiable space.
- Educate the public about living buildings.
- Design buildings that are truly beautiful and that elevate the human spirit.

Differing from the LEED rating system in one essential feature, the Living Building rating system has no points to acquire, only prerequisites. In other words, the Living Building is designed to be "eco-effective" and not merely "eco-efficient." It is a building that is positively good and not just "less bad." This concept presents a major challenge to architects and engineers, because it doesn't allow them to rest on their laurels until they have completely eliminated adverse environmental and social impacts of buildings.

Recent work in China, involving master planning by the internationally renowned firm Arup for a major new Dongtan Eco-City immediately west of Shanghai, shows that for a development to have a zero carbon footprint, working on individual buildings is not enough; attention must also be paid to common energy-using systems and to transportation energy use.[98]

● ● ●

Locally Sourced Materials

About 20 years ago in Italy, the "slow food" movement began, with an emphasis on the sourcing of local foods with higher nutritional content and a less rushed way to eat, certainly as a reaction to the American fast-food movement sweeping the globe. In the US, many estimates have food traveling an average of 1,200 miles from farm or fishery to table. Think about this the next time you buy Alaska Copper River salmon in Arizona or grapes from Chile in midwinter in Chicago.

In the same vein of concern about the energy and environmental costs of moving building materials vast distances, and the impact of this on local economies, the green building movement adopted the concept of locally sourced materials, for materials coming from less than 500 miles from a project site (as the crow flies). I call this the "slow building" movement, because it evinces a concern for buying something else than just the cheapest (or even the best) building material.

To encourage development of locally based "conservation economies" all over the country, the LEED certification system awards points for projects using at least 10% of the value of all building materials from sources within 500 miles (achieving 20% gets a project two LEED points toward certification). This means that products must have been extracted, harvested (or recovered in the case of salvaged materials) and processed within that radius. Common building materials such as concrete and wood generally qualify, but steel typically does not. (However, steel offers high levels of recycled content that wood and concrete do not, so this is a clear case of having to trade off among various beneficial environmental attributes.)

Consider concrete: nearly 95% of the weight of the material (excluding only the cement) comes from the local area, including water, aggregate and sand, so we can count 95% of the total cost to help meet this credit. If a project uses a lot of wood, some investigation may be needed to determine its source. In regions such as the Pacific Northwest, even softwood lumber from Canada might qualify for this credit. In the industrial heartland, the Midwest, even the iron ore in steel might qualify if it had been extracted in Minnesota.

Here are examples of some specific materials that could come from just about any locality or 500-mile radius, without traveling long distances: foundation piers; compost and mulch; concrete storm drains; masonry, pavers and hardscape materials; reclaimed lumber; wheatboard panels; most wood products, including laminated beams, cabinets, subflooring, roof decking, composite wood siding, engineered wood products and oriented strand board; millwork, both new and reclaimed and cellulose insulation. The list is seemingly endless.[99]

● ● ●

Low-flush Toilets

The current plumbing code requires toilets that use no more than 1.6 gallons per flush. This standard was adopted in 1992 (replacing the old limit of 3.5 gallons per flush) and has not changed in 15 years, even as water problems in the US are escalating. Responding to the growing concern over excessive water use in buildings, many companies have begun to offer even lower-flush toilets and urinals. Two good examples are dual-flush valves for tankless toilets (the kind you use in public places) and a dual-flush toilet with a 1.6 gallon flush for solid matter and a 0.8-gallon flush for liquid.

In January 2007 the US Environmental Protection Agency introduced a voluntary standard, with a 20% reduction in the 1992 levels, down to 1.28 gallons per flush. Manufacturers who meet this new standard can display the "Water Sense" logo on their products. One manufacturer estimated that a family of four could save 7,000 more gallons of water per year using a dual-flush toilet.[100]

Australia is suffering through a multi-year drought, said to be the worst in 100 years. Water levels in reservoirs for major cities such as Sydney had fallen to less than 30% of capacity, and the federal government is preparing to spend more than $10 billion (Australian) on a water desalination plant for the city. Watering restrictions are in force everywhere, with watering of gardens only allowed once or twice a week during the summer.

Because of the drought, dual-flush toilets have been mandatory in all Australian new homes since 2003. North American homes would be well served with similar requirements, as we get used to more permanent drought conditions and water shortages in the West and Southwest, brought about by population growth and global warming.

One of the major manufacturers of dual-flush toilets is an Australian company, Caroma. Living in a LEED Gold-certified apartment in Portland in 2005 and 2006, I used a Caroma dual-flush toilet, and I can attest that it worked just fine. In addition, with a large button for a big flush and a small button for a small flush, even a kid can figure it out. In most homes, you'd never know the difference in performance between the two, but you'd save about 40% of the water you would normally use for flushing, figuring three liquid flushes for every solid flush.

If you ever wondered how toilets are evaluated and rated for water use, the testing labs use thousands of flushes to evaluate their performance characteristics. You may be surprised, and perhaps feel enlightened, to know that soybean paste has approximately the same specific gravity and consistency as fecal matter. In 2005 a Canadian testing operation was reported to have imported 18,000 pounds of "non-food-grade" soybean paste (aka miso) from Japan, because it had to simulate hundreds of flushes for each toilet tested.[101] So, next time you drop into a local Japanese restaurant and are served miso soup, remember to smile.

Microturbines

Microturbines are a relatively new technology with significant applications in green building design. By using natural gas as a fuel (also diesel or propane), microturbines generate electricity and hot water, rather than just one or the other. In this way, about 80% of the energy value in the fuel is converted to useful work. Microturbines can range from 25 kilowatt (kW) output to 500 kW, have low emissions of nitrogen oxide, are about 20% to 30% efficient in producing electric power and can produce hot water at 120°F to 175°F, a range quite suitable for a number of uses, including swimming pools and service water.

Microturbines offer a number of potential advantages over other technologies for small-scale power generation. These include a small number of moving parts, compact size, light weight, greater efficiency, lower emissions, lower electricity costs and ability to use waste fuels such

Capstone Turbine Corporation

Microturbines at 60 kW power output are about the size of a large refrigerator. They are typically installed in group of five or more, to match demand for hot water and electricity in a building.

as biodiesel. They can be located on sites with limited space for power production, and waste-heat recovery can be used to achieve total system efficiencies of more than 80%.[102]

Microturbines typically come in 60 kW modules, about the size of a large refrigerator. By coming in small modules, it is easy to assemble a group of microturbines into an onsite power system and to match the electrical and thermal output to the building's demands. For example, the turbine's heat output can be used for water heating in a hospital or hotel, a facility type that requires lots of hot water on a 24/7 basis. If there is a swimming pool that gets a lot of use, any excess hot water can be used to heat the pool (which loses heat mainly through evaporation).

Other facilities that can benefit from microturbines include data centers, schools and colleges, food-processing or manufacturing plants, supermarkets and even sewage treatment plants.

The benefits of microturbines today are the same as those of cogeneration systems; they are cost-effective whenever there is a connected thermal load that uses heat most of the time. The electricity generated by the microturbines displaces the purchase of energy from a utility, at full retail rates; and the heat displaces natural gas that would have to be purchased otherwise just for a single purpose. In many cases, there is less air pollution and lower carbon dioxide emissions than from conventional generation.

● ● ●

Measurement and Verification Systems

How does a green building maintain its energy savings over the long haul? This is one of the critical questions in green building design, since there is plenty of evidence that building energy performance degrades over time. Systems wear out, and new building maintenance and operations people may fail to make necessary repairs, carry out preventive maintenance and generally fail to manage the building's energy-using systems as originally designed.

Green buildings are encouraged by LEED to take two simple measures to counteract this tendency toward energy-efficiency degradation. First, projects can gain a LEED point by developing a monitoring and verification plan, following established international protocols and then installing sensors that measure the actual performance of key energy-using systems such as chillers and boilers. The sensors are connected to the building automation system and provide information that allows engineers and building operators to pinpoint problem areas and fix them. Creating the plan and installing extra sensors are not that costly, typically $30,000 to $50,000, which is barely more than pocket change for a large office or residential building.

Additionally, building owners are encouraged to document systems training so that future operators can learn proper use and maintenance of the systems. The LEED for Existing Buildings standard encourages buildings to be re-commissioned every five years, so that energy performance can be maintained over the building's lifetime. For most institutional building owners, this is a smart thing to do since they're paying the bills.

The leading force in the US behind measurement and verification is the Federal Energy Management Program, which developed the International Performance Measurement and Verification Protocol, to identify and codify best practices techniques for verifying the energy performance of new buildings.

Think for a moment of the simple task of comparing this year's utility bill to last year's to determine if you're using more or less energy. What could affect the outcome? Weather is certainly a major variable: was this year colder or hotter than last year? Did the use of the building change, so that there were more or fewer occupants? Did the hours/days the building was occupied change significantly (for example, did someone put on a second shift)? Did someone put in a data center that uses a lot of power for servers, generating also a lot more waste heat that increased cooling demand? What about changes in lighting levels, occupancy sensors, ventilation levels and set-point temperatures for heating and cooling? Was preventive and remedial maintenance carried out? Were portions of the building vacant for any substantial period? How would *you* determine all this without a good plan and enough measuring and monitoring points to get accurate information? You can see the wisdom of planning ahead by following the LEED protocol for creating a plan, installing enough sensors and then collecting the data.

Native American and Native Canadian Ways of Living

If there is an underlying theme in green buildings that has a long history, it's that our contemporary civilization needs to learn the art of "living in place" for an extended period of time. For most North Americans, the Native American and Canadian way of life, with respect for the land, viewing the Earth as a Mother who gives all life, offers a way to live today to benefit seven generations into the far future. Most Americans would also agree that the attraction of this way of life is nostalgic at best, given our present urban society. Nevertheless, Aboriginal traditions exert a powerful pull on our psyche and have found expression in a number of elements of green buildings and green development. The idea of preserving open space and natural habitat is one way of honoring nature, while preserving natural elements in buildings is undoubtedly good for the psyche.

Another way Native American and Native Canadian approaches are being incorporated into buildings lies in climate-responsive design. Many architects are inspired by the Mesa Verde cliff dwellings in southwestern Colorado, where the overhanging cliffs protect against the harsh summer sun, high in the sky, while still allowing the lower-angle, warming winter sun to enter the homes. The adobe and stone buildings also stay cool in summer and warm in winter through their "thermal inertia," the ability to

© iStockphoto.com/Weldon Schloneger

Taos Pueblo, New Mexico, providing "homeland security" since 1492.

soak up heat during the day, for example, and release it slowly during the evening. Other Southwestern developments have been influential in local architectural styles. In 2006 I visited the Taos Pueblo in northern New Mexico that has several stories of apartments constructed of adobe bricks in two structures on both sides of a small river. There's a large plaza and a recorded population of nearly 4,500 people, but otherwise nothing remarkable to suggest that this site has been continuously occupied for almost 600 years. However, the adobe building style is endemic to New Mexico and has influenced building styles throughout the arid Southwest.

In the maritime Pacific Northwest, the Native American building style is the longhouse, a wooden structure large enough for an extended family, with plenty of access to nearby woods, rivers and the ocean — sources of abundant food, shelter and clothing. Preservation of rivers, riverbanks, public access to the beaches, marshland habitats, native plants and old-growth forests is very much part of the Northwest psyche, as are increasingly strong efforts to protect the varieties of wild salmon, the totem animal of that region.

● ● ●

Nature, Design with

Originally popularized by landscape architect Ian McHarg in the 1960s and still promoted by many landscape architects, "design with nature" starts with the land, geology, native plants and animal species, climate and water patterns, and then uses these dynamics to inform site planning. McHarg's premise was that ecology, the science of interrelationships between animals and plants, should be the basis for land planning and site design, at a regional scale as well as for individual sites. This approach helps to avoid stupid mistakes, such as building in areas prone to mudslides, earthquakes and floods.

"Nature always bats last" is an apt slogan for land development. We can look at the aftermath of Hurricane Katrina as a great example of how powerful natural forces can be, and while we sympathize with those displaced, killed or injured and economically hurt by Katrina, we can also see the futility of trying to rebuild a large city located in a place that will eventually become part of the Gulf of Mexico.

The green building movement addresses design with nature in a number of ways. First, we don't want development on sensitive ecological sites within 100 feet (or greater distance) of wetlands or rivers, on prime farmland or on land that is habitat for rare or endangered species. Second, we want land development to be more compact than the traditional, post-World War II suburban sprawl model. This means leaving more land in open space, even in greenfield developments. Ironically, developers are discovering that leaving open land in and around a development makes the remaining building sites even more valuable because future residents will have access to natural areas, trails and wetlands right outside their back doors, something that most people treasure. In Florida, where large tracts of swampland are unbuildable, developers have taken to touting the natural areas as "conservation easements" to attract homeowners and tenants interested in environmental preservation and willing to tolerate the occasional alligator encounter while jogging or walking.

When I sought a home in Tucson in 2006, I found a relatively new housing development where the natural desert had been left intact around the building lots, so that the local fauna (coyotes, bobcats, jackrabbits, chipmunks, the occasional javelina or peccary, quail, snakes, roadrunners and multitudes of birds) could coexist with the homes in perpetuity. For me, this is a very attractive way to live, even if I have to take extra precautions to protect my small dog.

Another quite different example of design with nature is the way landscape architects are introducing more natural water features into building projects. A leading exponent of finding out how water wants to flow in a park is a German designer, Herbert Dreiseitl, who speaks about how closely urban design and settlement patterns have been linked with water and its use, both functionally and artistically. As we are water creatures, one expression of biophilia in design is allowing water to be expressed in many ways, from flowing channels to water walls, to fountains and plazas, ponds and water-based microclimate cooling systems, stormwater collection and treatment with constructed wetlands.[103]

● ● ●

New Urbanism

New Urbanism is a movement launched in the early 1980s by planners and architects such as Andrés Duany and Elizabeth Plater-Zyberk in Miami. One of the earliest projects to demonstrate the principles of the New Urbanism was the village of Seaside in the Florida panhandle, near Ft. Walton

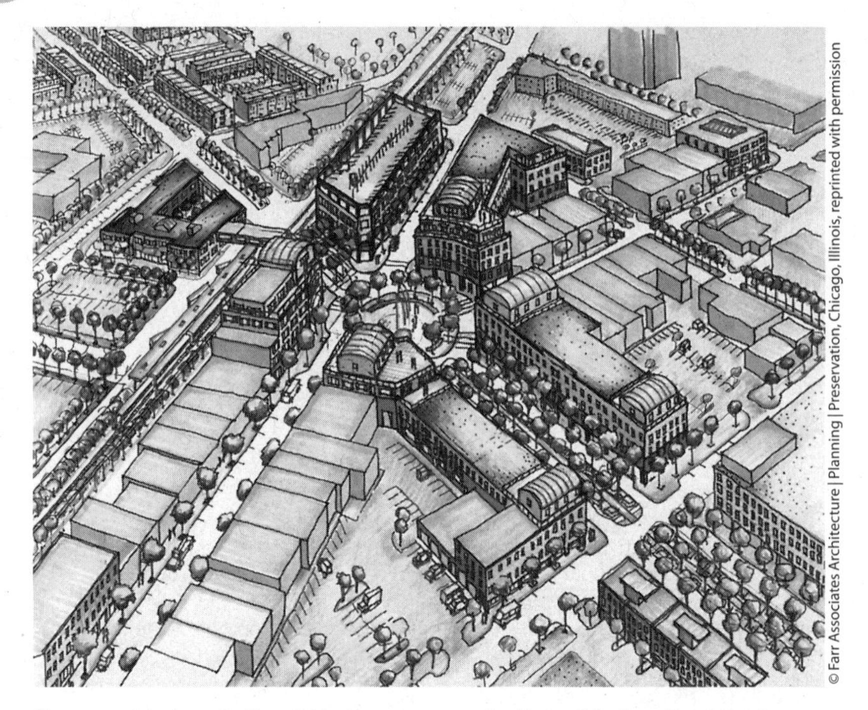

Conceptual design of a New Urbanist town center for Normal, IL, from Farr Associates, Chicago.

Beach. These developments are compact neighborhoods, often with densities of eight to ten units per acre, with front porches so neighbors can actually see and talk to each other. They are also very walkable, with key amenities such as a grocery store and a transit or bus stop within a few hundred yards. A typical New Urbanist plan, in this case for Normal, Illinois, shows how to design these elements.

The New Urbanism is often linked to a related movement toward Transit-Oriented Development: building homes, offices and commercial development at or near light-rail stations or other major transportation hubs, to make it easier for people to avoid using the single-occupant vehicle to get to/from work, home, recreation, etc. A growing body of research indicates that people are healthier in places where they can walk or bike to basic services instead of using a car for every errand. Think of your own experiences, living, working or visiting great cities like New York, San Francisco, Boston or Chicago where you can (and do) walk or take transit to most places.

In new communities, New Urbanism today is particularly expressed in several key concepts:

- Connecting communities by locating shops and basic everyday needs within walking distance.
- Neighborhood location, so that people can walk to transit and also walk separately from roadways.
- Traffic calming — using various methods to slow down cars — to make streets safer and street life more viable.
- Land-use patterns that respect natural drainage contours, wildlife corridors, etc. by increasing density in built-up areas to allow for more open space in a development.

Onsite Sewage Treatment

We live in a "flush without fear" society, in which most of us just flush away and presume that "they" will take care of it somewhere downstream. Of course, for most of us, our downstream is someone else's upstream! Nevertheless, for the most part, this is a reasonable presumption, since modern sewage treatment systems have been perfected over the past 100 years, and flush toilets have been around since Thomas Crapper first popularized them in London in the 19th century.[104] However, a basic principle of sustainable design is that there is no "away" place to dispose of waste, since everything is interconnected.

Urban sanitation systems, with secondary and even tertiary sewage treatment, still burden many rivers and coastal waters with excessive amounts of sewage sludge and industrial waste, even in modern societies. In older cities, combined sewer overflows during rainy periods still pollute local rivers. Therefore, green building designers are increasingly considering treating sewage onsite, rather than sending it offsite. Engineers are developing new ways to treat sewage in confined spaces, such as building basements, using both aerobic and anaerobic digestion processes as well as a final "polish" with ultraviolet radiation to kill any remaining microorganisms.

In one project the savings from not paying fees to connect a project to the local sewer system paid for the capital cost of having a third-party firm design and construct a system to serve 1,000 people per day in a 16-story building. The same firm will operate the system and charge the project the same amount it would have paid to the local sewer utility. At no cost to itself, the project will recycle five million gallons per year of sewage for use in flushing toilets and irrigation.[105]

Other systems use land disposal by filtering waste through constructed wetlands or through treatment in a series of ponds or tanks in more confined spaces such as greenhouses, a system first pioneered in the 1970s by John and Nancy Todd of the New Alchemy Institute. In a "living system" greenhouse, the nutrients in sewage feed algae and water plants that in turn nourish fish. The final effluent is quite clean and can be reused in various ways. Using treated wastewater is nothing new. Many cities use reclaimed wastewater for irrigating golf courses and other spaces such as median strips and parks. Properly applied, it is a beneficial way to save water and reuse the nutrients present in wastewater.

For some green building projects, there may be a significant financial benefit from onsite sewage treatment, namely, that it helps reduce water use dramatically and often saves considerable money upfront that would be paid for a municipal sewer connection. It can also reduce utility costs, since water supply charges in many places also include significant charges for sewage treatment. These savings can be enough to pay for the cost of constructing and maintaining an onsite sewage treatment system.

And, of course, why not use treated sewage for flushing toilets, since "black water" is about 98.5% water anyway? That helps close the water cycle, an essential element in a truly sustainable building. Other uses for treated sewage in a building can include washing of roadways and sidewalks, landscape irrigation and cooling tower "makeup" water (to replace water lost by evaporation and back-flushing).

Operations and Maintenance Practices

Environmentally beneficial operations and maintenance (O & M) practices are just beginning to come into use, spurred by the LEED for Existing Buildings (LEED-EB) rating system. Building owners and managers are beginning to examine their use of chemical fertilizers, pesticides, herbicides and other toxic substances, to eliminate those that pose significant hazards to people, animals and the environment. Large organizations with sustainability mission statements are beginning to see the benefits of changing practices, in terms of cost savings and reduced exposure of their people to harmful substances.

JohnsonDiversey's global headquarters in Sturtevant, Wisconsin, is a three-story, 277,000-square-foot mixed-use facility, 70% office and 30% labs. It received a LEED-EB Gold designation in 2004. At an implementation cost of $74,000, or $0.27 per square foot, the company has documented annual net savings of $137,000 ($0.49 per square foot), with a life-cycle net present value of $1.35 million, a return (in today's dollars) of almost 20 to 1 on the initial investment! Energy savings were documented at $90,000 per year, potable water use was reduced by two to four million gallons per year, and more than 50% of site-generated solid waste was documented as recycled. The program has renewed the company's focus on using integrated pest management, training cleaning workers and developing an integrated program of green cleaning aligned with LEED requirements.[106]

In a more urban setting, in 2003, Craig Sheehy of Thomas Properties Group in Sacramento tackled the 25-story, 950,000-square-foot headquarters of the California Environmental Protection Agency (Cal-EPA), which originally opened in 2000. The Cal-EPA building was awarded the first LEED-EB Platinum certification in 2004. Through the program, it gained 34% in energy efficiency, diverted 200 tons or more of waste from landfills each year and added $12 million in asset value (it is privately owned and leased to Cal-EPA). Over ten tons per year of food waste was vermicomposted and bagged as "Bureau Crap" for home garden use. From an investment of $500,000, more than $600,000 in annual savings was realized. This shows what LEED-EB can do with even a relatively new building.[107]

In 2006 Adobe Systems, Inc., a software maker in San Jose, California, certified three of their headquarters buildings at the LEED-EB Platinum level. Since 2001 Adobe had invested approximately $1.1 million for energy and environmental retrofits in the three towers, saving approximately $728,000. Over this period, Adobe reduced electricity use 35%, natural gas use 41%, domestic water use 22% and irrigation water use 75%. Adobe now recycles or composts up to 85% of solid waste. Through energy savings and purchase of green power, Adobe reduced its pollutant emissions by 26%.Working with facilities manager Cushman & Wakefield, Adobe installed drought-tolerant landscaping, with an irrigation system linked to local weather stations, and added building sensors to monitor interior carbon monoxide levels and adjust the operation of exhaust fans accordingly. Adobe also increased its use of outdoor (fresh) air for ventilation and cooling and enhanced the overall maintenance of its air systems, resulting in better indoor air quality.[108]

Ozone-layer Protection

The Montreal Protocol came into force in 1989. It bans the production and use of chemicals that have been shown to damage the ozone layer, a concern first raised by scientists in the 1970s. Without the protective ozone layer, we all would get skin cancer! Discovery of a persistent ozone hole over Antarctica in 1984 by British scientists highlighted the damage to the ozone layer and heightened public concern over the chemicals that were believed to be causing the phenomenon.[109]

These damaging chemicals include commonly used refrigerants, such as Freon and chlorinated fluorocarbons (CFCs). With the signing of the Montreal Protocol, chemical companies began research in earnest to come up with substitutes that would be as efficient as conventional refrigerants in cooling buildings (and everything else) without having any environmentally harmful side effects. The issue is simple: less-efficient refrigerants mean more electrical energy is needed to accomplish the same amount of cooling (engineers express it in terms of kilowatts of electricity per ton of cooling); therefore, more carbon dioxide emissions occur from fossil-fuel-fired power plants (the dominant US mode of producing electrical power), and more global warming occurs.

According to one expert, global warming and ozone depletion are both undesirable, and the two effects are negatively synergistic. Ozone depletion aggravates global warming, and global warming aggravates ozone depletion.[110]

The LEED system bans CFC use in "base building" cooling and refrigeration systems (excluding water coolers and small refrigerators) and also bans CFCs, HFCs (hydrofluorocarbons) and halons in fire suppression systems.

Better living through chemistry! What we have discovered this decade is that chemicals with high ozone-depletion potential generally have low global-warming potential *and* vice versa. So, the attempt to prevent ozone layer damage might inadvertently lead to higher levels of global warming. Recognizing this conundrum, the LEED system was modified a few years ago to allow some hydrochlorofluorocarbons (HCFCs), such as HCFC-22 and HCFC-123, in building refrigeration systems, so long as they have relatively low global-warming potential. Under the Montreal Protocol, HCFCs are not scheduled for a complete phase-out in developed countries until 2030.

When older buildings are being renovated, it's important to replace older air-conditioning and refrigeration systems with modern chillers and refrigerators using HCFCs that are relatively benign both in terms of global warming and ozone depletion. In new buildings, why not design them to operate whenever possible without mechanical cooling systems? In the maritime Pacific Northwest, relatively low humidity and cool temperatures allow properly designed buildings to operate on 100% outside air a good part of the year. In other areas of the country, particularly along the coasts and in higher elevations, passive solar design, operable windows, radiant cooling and natural ventilation strategies will allow systems to operate without mechanical cooling for months each year.

Paints, Low-VOC

In 2006 I moved into a home that needed repainting. Since my wife is a "miner's canary," in terms of her sensitivity to all chemical emissions, we went in search of paint that wouldn't leave a strong odor. After some looking, we found an "ecological" paint from a major manufacturer with only 3 grams per liter of Volatile Organic Compounds (VOCs), versus 127 grams per liter for their conventional paint. Thinking that would be just fine, we added the color we wanted and took it home. Guess what? The color added so many volatile solvents that the paint still bothered my wife significantly.

Fortunately there are options for buying low-VOC natural paints. In a city with an ecologically focused home-improvement store, you can get expert consultation on low-VOC paints. One unique approach to paint selection is at the Ecohome Improvement store in Berkeley, California. There, you can sit around a "paint bar" and a knowledgeable "paint-tender" will show you the choices.

Another approach is to choose an entirely new way to make paint. Green Planet Paints is headed by Meredith Aronson, an entrepreneur in southern Arizona with a Ph.D. in chemistry. She is beginning to hit the market with paints made from clay, marble, mineral pigments and a soy-based resin that makes the surfaces washable, all based on ancient Mayan techniques and ingredients. These paints have no VOCs at all. Of her more natural paints, Aronson says, "The environmental footprint of even 'zero-

Ecohome Improvement, Berkeley

Paint bar at Ecohome Improvement in Berkeley, CA, designed and built by Salvage. The paint bar is made from Vetrazzo, a recycled glass countertop, and reclaimed old-growth redwood from a dismantled Bay Area water tank.

VOC' paint can include all kinds of synthetic materials to control flow, skinning, settling, etc. that ultimately don't support a vision of sustainability and goodness for the environment."[111]

In larger commercial settings, there are of course many options, and the LEED system has very defined rules for limiting VOCs in paints and coatings below threshold levels. These limits, 50 grams per liter for flat and 150 grams per liter for non-flat interior paints, are set by the Green Seal standard, GS-11.[111] They are still a far cry, however, from "zero-VOC" paints that must contain no more than 5 grams per liter.

Paradigm Shift

Green buildings represent a major paradigm shift in architecture, engineering, construction and development. Instead of evaluating buildings on aesthetics or economics, we are now requiring that they be assessed against a set of criteria such as those which make up the LEED rating system, which look at energy, environment and health criteria. Make no mistake, this is a significant challenge to conventional wisdom and business as usual in the design and development business.

What is a paradigm? It is a dominant way of seeing the world, so embedded in our thinking that we don't/can't even consider alternatives.[112] A few examples are familiar to everyone: 500 years ago, proposing that the Earth was round and revolved around the sun ran counter to both religious and scientific dogma. Espousing this new paradigm was tantamount to religious heresy (think of Galileo's imprisonment in the early 1600s), yet the facts were increasingly at odds with the old paradigm. Indeed, by the time Magellan's crew returned from an around-the-world voyage in 1522, traveling only in one direction, the roundness of the Earth was no longer an issue for medieval society. Within 100 years, telescopes and careful observations by astronomers began to establish planetary rotation around the sun as fact.[113] To suggest otherwise today would be ridiculous.

In 1905 Einstein's remarkable special theory of relativity challenged more than 200 years of Newtonian physics that held that matter and energy were separate entities. By linking them together, Einstein overthrew the prevailing scientific orthodoxy and paved the way for a revolution in physics, as well as for atomic and hydrogen bombs, the development of nuclear power and nuclear weapons proliferation. In 1917 Einstein's general theory of relativity blew away the notion that space and time were separate, paving the way for theories of an ever-expanding universe. In just a few years, this theory was confirmed by observation.

New research in medicine, neuroscience and other disciplines is attacking dominant paradigms about mind, body and consciousness that have ruled for hundreds of years since French philosopher René Descartes first postulated, "I think, therefore I am." Global warming and sustainability concerns may be the forces that lead to a paradigm shift in our current dominant understanding of the world, one in which both unlimited economic growth and growing fossil fuel use are without consequence!

Given the major issues of global warming, loss of biodiversity, water and energy shortages in much of the world, green buildings represent the beginning of a paradigm shift toward truly sustainable design and development, restoration of ecosystem functioning and zero-net-energy urban settlements. As with most paradigm shifts, it's hard to see how we're going to get such results with our current systems of budgeting, planning, design, construction and operation of buildings and facilities. Yet, as scientist Donella Meadows famously observed in the 1990s, only a paradigm shift can make a significant change in complex human systems.[114]

● ● ●

Passive Solar Design

Passive solar design refers to a number of intelligent building design techniques that reduce or eliminate the use of fossil fuels and electricity for heating, cooling and lighting buildings (during the day). The modern version of this traditional approach to building design was developed beginning in the 1970s and applied to a wide variety of building types throughout the US, with a focus on the West and Southwest.[115] The term "passive solar design" was chosen to contrast with the more prevalent — and far more expensive — active solar systems that used expensive copper or aluminum rooftop collectors and lots of fans, pumps and controls to heat and cool spaces. The idea behind passive solar design was to incorporate sunlight and natural ventilation into the basic design of the building, minimizing the need for mechanical systems. In many of the hot, arid climate zones of the US, this is an excellent design strategy. In hot, humid zones, more focus needs to be given to ventilation and less to heating.

In 1980 I built a passive solar adobe home in the San Francisco Bay area. In a climate that typically ranged from cool winter nights (in the 30s) to hot summer days (in the low 100s), the three-bedroom post-and-beam design featured a heavy-mass structure (adobe clay), plenty of south-facing glass, a well-insulated roof and abundant cross ventilation between low north-facing windows and high south-facing clerestory windows. The long axis of the home was east-west, with small windows on those façades. The south-facing windows were shaded with overhangs to prevent the high summer sun from entering the house, but did allow the lower-angle winter sun to shine directly on the mass of the structure, heating the home not only during the day, but also well into the evening. The home had abundant daylighting, and temperatures in the space stayed 20°F above outside levels in winter and 20°F below outdoors in summer, comfortable almost year-round. Only a small woodstove in the living room was used for supplemental heating, and no air conditioning was used. Even on hot summer days in the Bay area, there are cooling breezes most evenings that lower temperatures quickly.

Applied to commercial buildings, passive solar design looks for opportunities to orient buildings with a long axis east-west, to have a more rectangular than square structure (called "massing" by architects) and to provide shading and overhangs on the south and west faces using a window design that allows for abundant daylighting with minimal glare.[116] In smaller buildings, it makes design sense to have direct solar penetration into a space, such as an atrium, to warm it in winter. Such buildings can also use radiant heating systems to supplement solar heating in winter, and many use concrete or other forms of thermal mass to absorb heat during the day in summer, radiating back slowly at night when people have left the building. Some approaches also use a strategy known as "night-flush cooling" in which fans slowly circulate cooler night air through the building (and exhaust warmer air) to lower temperatures to comfortable levels before morning. This approach saves a lot of energy compared with conventional air conditioning.

● ● ●

Permeable Pavement

Permeable pavement is one of those great innovations about which everyone says, "Why didn't they come up with this sooner?" In parking lots, its purpose is to allow water to infiltrate directly into the ground, instead of running off into the streets, storm drains and eventually into rivers and oceans, polluting them with oil and grease. EPA states that "used oil from one oil change can contaminate 1 million gallons of fresh water."[117] The main source of this runoff is the incredible amount of impervious surface area we have put on the land. As Joni Mitchell sang in a 1960s environmental lament, "They paved paradise and put up a parking lot."

One of the main sources of ocean pollution is so-called non-point source (NPS) runoff from streets and parking lots, carrying oil and grease (and other contaminants) from parking lots into coastal waters. The most common NPS pollutants are sediment and nutrients, washing into water bodies from agricultural land, small and medium-sized animal feeding operations, construction sites and other areas of disturbance; other common NPS pollutants include pesticides, pathogens (bacteria and viruses), salts, oil, grease, toxic chemicals and heavy metals.[117]

The trick in permeable paving is to design the parking lot right in the first place to allow for subsurface water infiltration. The Port of Portland installed a 35-acre porous asphalt pavement to allow stormwater to drain through a paved surface and recharge groundwater. Incorporating a series of bioswales (planted ditches) and natural vegetation allowed all of the stormwater from the expansion to infiltrate the ground.[118]

There are good environmental reasons to let rainfall take its natural course and run off into streams and lakes, but the increase in flooding from urban development and the polluted nature of the runoff make the argument stronger for keeping as much onsite as possible. In many larger metropolitan areas along the coasts, authorities are beginning to require that large developments design their landscaping and parking areas to hold runoff onsite, either in detention/retention ponds, bioswales or similar devices. In these cases, permeable pavement might be an excellent complementary technology that would allow a developer to reduce the size of other drainage elements.

Permeable paving can include a variety of techniques. Porous asphalt contains conventional asphalt with small aggregate omitted from the mixture. Placed under the porous asphalt surface is a base of further single-sized aggregate. Porous concrete, like porous asphalt, can bear frequent traffic and is universally accessible. Single-sized aggregate (usually seen in gravel parking lots) without any binder is the most permeable paving material in existence — and the least expensive. Although it can be used only in very low-traffic settings such as seldom-used parking stalls, its potential cumulative effect is great. Porous turf is sometimes used for occasional parking like that at churches and stadiums. Living turf transpires water, counteracts the heat-island effect with what appears to be a green, open lawn. Open-jointed blocks are concrete or stone units with open, permeable spaces between the units. They give an architectural appearance and can bear surprisingly heavy traffic.[119]

● ● ●

Photovoltaics

Photovoltaic (PV) solar electric systems ("photo-voltaics" means electricity from light, especially sunlight) have been around since the 1950s and have been used extensively in world's space programs to power satellites, the International Space Station and other spacecraft needing power for operations. Using semiconductor-grade silicon, solar cells convert the energy in sunlight to electricity, typically at efficiencies of 5% to 12%, or with power output of about 5 to 12 watts per square foot. Sunlight is a very diffuse energy, falling on the Earth at the rate of about one kilowatt (the input for ten 100-watt bulbs) per square meter. Electricity is a very concentrated form of energy, so it's not surprising that it takes considerable collection area to make any sizable amount of electricity from the sun.

Nevertheless, solar electricity is rapidly gaining popularity around the world. For many years, solar power in the US has been used by thousands of homes to become "grid-independent" and by larger projects to make a quite visible statement about their use of solar energy. Despite our abundant solar resources, especially in the Sunbelt states and the West, the US government has never put much of a priority on commercializing photovoltaics.

The 2005 Energy Policy Act (EPACT) created a 30% federal tax credit (residential credit limit is $2,000, commercial is unlimited) for solar electric installations, currently set to expire at the end of 2008. Many utilities also have solar electric promotional programs, and some state governments such as Oregon and Arizona have additional tax credits. In 2007 the California Public Utilities Commission enacted a $2.2 billion "million solar roofs" program that requires the state's investor-owned electric utilities to offer incentives in the range of $2,800 per kilowatt for photovoltaics.[120]

The largest drawback to increased use of photovoltaics has been cost. Today's large commercial systems install at about $6,000 to $7,500 per kilowatt, while consumer systems can cost $7,500 to $10,000 per kilowatt. What does this mean in terms of delivered value for electricity? Assuming

p

that PV systems last 15 years (which they do), a simple amortization would mean an annual cost of $400 to $500 per kilowatt. PV systems will produce 1,200 to 1,800 kilowatt-hours per year, per installed kilowatt. So, to get 1,200 to 1,800 kilowatt-hours per year, you would have to spend $400 to $500. Even at retail rates (a program known as "net metering" or "running your meter backward") of 8 to 10 cents per kilowatt-hour, you would earn $96 to $180 per year, at a cost of $400 to $500, not exactly a paying proposition. If electric rates are 15 cents per kilowatt-hour, then the annual return would be $180 to $270.

But now assume that government incentives lower the cost of PV systems by 50%, so that your annual cost is only $200 to $250. You're still not making money. Are there still reasons to do this? Yes, there may be. For example, what is it worth to have a supply of electricity that is independent

Kettle Foods

To celebrate its 25th anniversary, Kettle Foods partnered with the Energy Trust of Oregon to install one of the largest solar energy arrays in the Pacific Northwest. Using more than 600 solar panels set on roof-mounted racks, the plant now generates 120,000 kWh of electricity per year. That's enough power to make 250,000 bags of chips and reduce annual carbon dioxide emissions by 65 tons.

of the electrical grid? Or that is fixed in price today, no matter what future utility prices may be? Or that allows you to make a statement that you are producing domestic renewable energy without any environmental impacts (except those from manufacturing semiconductor-grade silicon)? What if your roof could be made from PV-powered shingles or tiles, so that when you had to replace your roof, you could combine it with a PV system at a lower total cost? What if a partnership wanted to lease your roof (for $1 per year), install the system, take the various tax benefits and you only had to pay for the electricity produced?[121]

Here's how to calculate the cost of PV-generated power:

1. Cost of installation, less the value of all incentives, expressed as $ per kilowatt-peak (power rating).
2. Power generated: the typical US range is 1,200 to 1,800 kilowatt-hours per year, per kilowatt-peak power rating.
3. Value of power generated: typically your retail rate, 8 cents to 12 cents per kilowatt-hour.
4. System lifetime: assume 20 years.
5. Net Present Value (NPV) Factor: you'll have to figure a way to discount the value of future electricity, but figure a 5% discount rate (like a 20-year government bond); over 20 years, the value of getting $1 each year at 5% is worth $12.46 today. Remember this is pretty risk-free and typically generates a non-taxable return for a residence (but not for a business).

So, here's an easy formula to use:

Annual kilowatt hours produced	×	Your electric rate	×	NPV Factor	=	Value of PV electricity

In our example, the value equals:

1,500 kWh × $0.10 × 12.46 = $1,869

This formula tells you that could pay $1,869 (per kilowatt-peak) for the system and still generate a 5%, 20-year return, at 10 cents per kilowatt hour. If your retail rate is 15 cents, then you could pay up to $2,803 for the same return. If you think power costs will increase faster than the rate of inflation, you could increase the price you're willing to pay, after taking all incentives into account.

p

Platinum Buildings

If green buildings are the goal, then a lot of people shoot for the highest ranking possible, which is LEED-Platinum. Without exactly knowing what it takes, many building owners and design teams begin their green building project by proudly proclaiming a goal of LEED Platinum. Usually, rather quickly they find out there is more to making a project "super green" than just declaring good intentions.

As of early 2007, there were fewer than 30 LEED Platinum new buildings in the US in all four USGBC rating systems — about 4% of the total number of certified projects. Even among those buildings, there were some that barely made it over the bar (52 points out of 69 possible in the LEED-NC rating system), while others achieved as many as 60 points. (It's well-nigh impossible to get all points in any of the rating systems for a single project.)

Anton Grassl

Genzyme Center, Cambridge, MA. Behnisch Architects/ Architekten, with House & Robertson Architects and Next Phase Studios.

Gerding Edlen Development

Center for Health and Healing at Oregon Health & Science University, Portland, OR, designed by GBD Architects.

To date the highest point total belongs to the 110,000-square-foot (excluding parking) corporate campus renovation for Alberici Constructors, Inc., in St. Louis, Missouri, completed in 2005. In this building, a 50-year-old manufacturing facility on a 13-acre brownfield site was renovated into a modern two-story office building, with solar thermal panels for hot water on the roof and a 65-kilowatt wind turbine onsite for additional renewable energy.

At the end of 2006, the largest Platinum-certified building was the 344,000-square-foot, 12-story Genzyme Center corporate headquarters in Cambridge, Massachusetts, completed in 2003. Its building envelope is a high-performance curtain-wall glazing system with operable windows on all 12 floors. More than 32% of the exterior envelope is a ventilated double-facade that blocks solar gains in summer and captures solar gains in winter. Steam from a nearby power plant is used for central heating and cooling. The project is owned and operated by Lyme Properties, with Genzyme as the major tenant.[122] Including 20 kW of photovoltaics, overall energy use is projected at 41% less than in a conventional building.

In 2007 the Oregon Health and Science University's new Center for

p

Debbie Franke

Alberici Headquarters, St. Louis, MO, designed by Mackey Mitchell Associates.

Health and Healing, a 412,000-square-foot mixed-use medical office, lab and classroom building became the largest Platinum-certified project in the world. This project combines 60 kilowatts of building-integrated photovoltaics, a large site-built solar collector for water heating, a 300-kilowatt microturbine plant, an onsite sewage treatment plant, an extensive green roof and 100% recovery of all rainwater for reuse. Occupied in the fall of 2006, it is projected to save more than 60% of the energy use of a similar conventional building and more than 50% of the water use. Total cost premium was reported at 1%, net of all incentives.

There are also LEED Platinum projects in Dubai (United Arab Emirates), China and India. In 2007 and 2008 we expect at least 50 more projects, recently completed or currently under construction, including one as large as one million square feet, to receive LEED Platinum designations in one of the four established LEED rating systems.

Post-occupancy Evaluation

One of the key principles in sustainable design is to have a feedback loop, a process that helps organizations and individuals to learn from their decisions and to make better choices in the future. The key feedback loop in green building design is called post-occupancy evaluations (POE), in which someone goes back after a building has been completed to see if energy and water use are meeting projections, whether the indoor air quality is as predicted, whether the building operators are running the monitoring and control systems properly and so on. A key element in a POE is a survey of occupant comfort because so much of the resulting productivity gains expected from green buildings have to do with daylighting, lighting, ventilation quality and thermal comfort.

Developed in England, POEs have been slow to catch on in the US, primarily because no one has provided a budget for such work, and because most of the design and construction team has moved on to other projects. So it's left to the building operators and managers, many of whom might be under contract to the building owner, to make the building work.

In some ways the lack of interest in POEs represents a huge black eye for both architects and engineers. For architects, it indicates a lack of professional interest in how their designs actually work for the people in the building, beyond perhaps a few anecdotes. This may be because user groups are typically not well incorporated into the design process and have little input into important design decisions that affect the future occupancy of the building. Beyond commissioning the building at the end of construction, engineers seldom return to assess results. Perhaps it's because they are afraid of getting sued for buildings that don't perform as designed. If green buildings are to realize their full potential, there has to be money in every building budget to assess results and feedback to future designs.

A recent example of a successful POE was performed for Portland State University's (PSU) Stephen E. Epler Hall, a 130-unit dormitory that was awarded a LEED Silver designation in 2003. A master's degree candidate at PSU, Cathy Turner, did a POE for this project. The result evoked widespread interest in the Pacific Northwest in performing a similar assessment for other projects.[123] If one thinks about it, there are hundreds of graduate students in various sustainability-oriented degree programs who would welcome the task of performing POEs on behalf of building owners and project design and construction teams. All that's needed is a little funding for their time and a solid connection to their degree programs.

A study by the Center for the Built Environment at the University of California, Berkeley, compared 21 LEED-certified and other green buildings with more than 160 conventional buildings. The responses to more than 33,000 questionnaires showed that green buildings had higher occupant satisfaction overall (statistically significant difference) and specifically higher satisfaction with indoor air quality (average satisfaction in

80th percentile of users surveyed) and thermal comfort (average satisfaction in the 90th percentile).[124] This demonstrates the value of green building design properly executed to generate two key benefits: human health and worker productivity.

Productivity

Productivity gains are one of the major business case benefits of green buildings. Why is productivity so important in justifying green buildings? Consider typical annual building operations costs for people (salary and benefits), rent and energy. For example, a $60,000 per-year employee (salary and benefits) in an average space of 200 square feet will cost $300 per year per square foot. For commercial buildings, people costs are about 10 to 20 times greater than rent ($15 to $30 per square foot per year), which is in turn about 10 times greater than energy (about $1.50 to $3 per square foot per year). This result does not say that saving energy is *not* important, but rather that even a 1% gain in worker productivity will offset the entire annual energy bill.

Productivity gains from mixed-mode conditioning and natural ventilation systems.

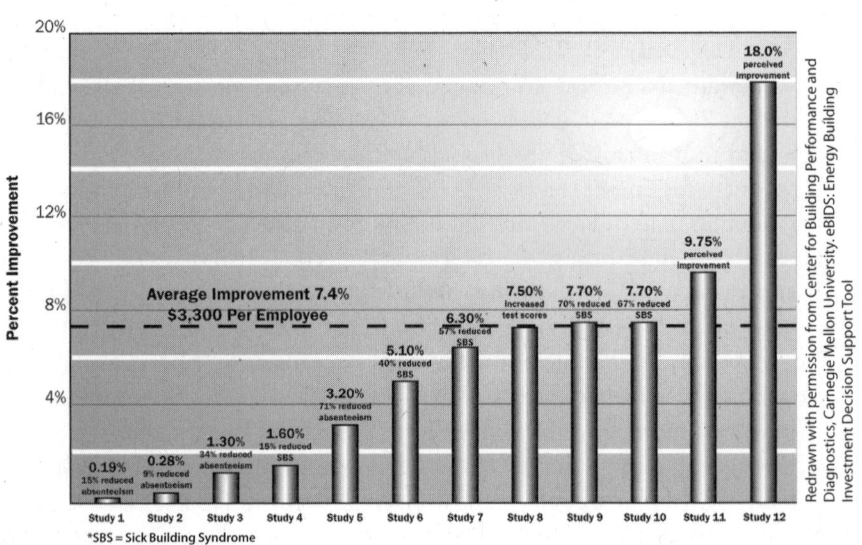

Moreover, a 5% to 10% gain in productivity will pay for the entire rent on a building. So, if a green building costs 5% more than a conventional building, but has daylighting, views of the outdoors and healthy indoor air, those features will likely lead to a productivity gain of 3% to 5% or more — that has a value of $9 to $15 per square foot in the first year! In other words, for a corporate or institutional owner who can reap the benefit of the investment, the first-year return on investments is more than 100%. If the funds are available, that return makes the investment a no-brainer. For this reason alone, green building design more than pays for itself, even if there are higher costs. If a company can realize a 10% productivity gain from a green building, it pays to build a brand-new building for employees!

Furthermore, there are dozens of studies that link higher productivity to a number of the building's green features. These buildings are also linked to improvements in illness and absenteeism among employees. If employees are healthier and on the job more often, productivity gains are a direct result. A number of academic studies[125] show gains in productivity from the use of mixed-mode and natural ventilation systems that average 7.4%.

q

Question Authority

Green buildings represent a challenge not only to the authority of architects and engineers, but also to the codes and standards prevailing in the building industry. In many cases, building code officials in cities and counties are unaware of new technologies, putting the burden on the design professional to make a case for evaluating a new idea on a "performance" vs. code-prescriptive basis. In a performance evaluation, the proposed design is assessed to see if it provides the same level of protection of public health and safety as the prescriptive standard. This may occur with such simple technologies as water-free urinals or with complex approaches to building fire protection and energy efficiency.

As an example, in 2002 one engineering design firm in Portland wanted to harvest rainwater from a building roof and reuse it for toilet flushing in a college dormitory, something the students favored. The design included a water treatment system using sand filters and ultraviolet light to destroy pathogens, so the finished product was basically safe enough to drink. As the first such project in the city, the code officials were a bit nervous. So they only allowed the first-floor public restrooms to have the recycled water, not the dorm rooms! And, they made the project put a sign over each toilet, proclaiming "Rainwater — Do Not Drink!" Since the toilets had no tank, but only a valve flush, this seemed a bit much!

Progress occurs in slow increments; these same code officials relaxed their restrictions with each successive project, so that in 2007 rainwater harvesting is now part of the tool-kit of engineers in the Portland area. Ironically, Oregon still prohibited water-free urinals at the end of 2006, largely owing to opposition from union plumbers, so efforts to conserve water in building projects there still face obstacles.

The point is that green buildings often butt up against a well-entrenched system for designing, constructing and operating a building. That system needs to change dramatically if we are to achieve the full benefits of green buildings. If even simple systems like rainwater harvesting take several years to achieve acceptance in each jurisdiction, how will we ever be able to make green buildings deliver, in a timely manner, the energy and water savings we all agree they need? How will dramatically different new technologies gain acceptance and critical mass for marketplace success if the battles have to be fought one project, one city and one contractor at a time? Local officials need to work closely with architects, engineers, contractors and the building products industry to make needed changes in conventional practices.

One final note: often the authority that needs to be questioned is not external, but more subtle, namely, the internal pressures on architects and engineers to be far more conservative in their designs than required by circumstances. This is the pervasive authority of rules of thumb, sales engineers, engineering handbooks and the collective experience of more senior architects or engineers in a firm. Their experience tells them, for example, to oversize HVAC systems to avoid future complaints from building occupants or the possibility of lawsuits over the adequacy of building design.

Radiant Heating and Cooling

We're so used to overhead forced-air ventilation, heating and cooling systems that we neglect to design buildings with other technologies for comfort. The figure below shows the four key variables for human comfort: air temperature, relative humidity, air movement and radiative temperature (the temperature of surfaces). Think of your own experience standing next to a cool brick or concrete wall on a warm day but not feeling uncomfortable because of the radiant-cooling effect of the wall. Or, consider a masonry or brick stove (the old "Russian" stove) radiating heat for hours even after a fire is out.

In commercial buildings, radiant cooling strategies are getting a new look, through the introduction of "chilled beams" (fins with water tubing that can circulate cool water, providing a surface that appears cool to our bodies even with warmer-than-normal air temperatures). Radiant floors can also be used for heating in cooler climates or for cooling in buildings that require year-round air conditioning. They can also be used in assembly spaces (such as an atrium of a larger building) when exact temperature controls are not required. One LEED Gold-certified renovation project

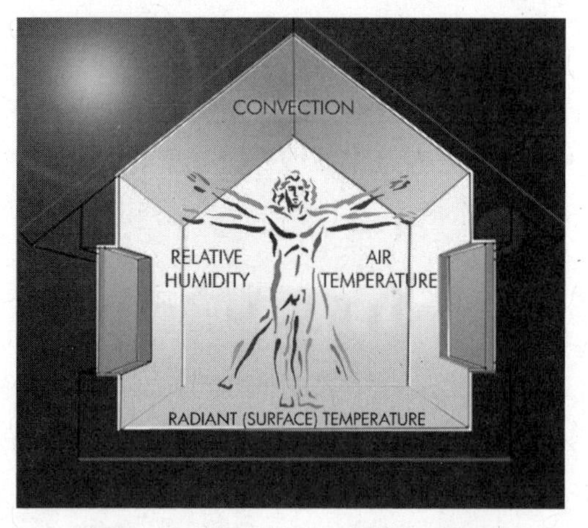

Determinants of Comfort

Comfort in a room, office or enclosed space is based on the subtle interaction between many variables, including humidity, air temperature, air movement and the temperatures of the surfaces your body "sees" in the space.

even poured a thin concrete slab on top of an old wooden floor to install radiant heating tubing.

The benefit: often the size of other building HVAC systems can be reduced, saving money on initial costs, since less air movement is required for comfort (and since most fan systems are sized for cooling, not just for ventilation). In addition, air temperatures can vary by 3°F to 4°F in either direction outside of a normal range, without occupants feeling uncomfortable. With higher air temperatures, less energy is required for cooling and less fan power is used for air movement. Elementary physics tells us that hot or cold water contains a lot more heat (heat capacity) than the same volume of air, and that far less energy is required to pump water than to blow air to get the same comfort effect. Why fight nature?

Rainwater Reclamation/Reuse

One of my favorite green building technologies is rainwater harvesting: the capture, treatment and use of rainwater for uses inside the building such as toilet flushing and cooling-tower makeup water (to replace water lost by evaporation and back-flushing). This is such a simple and obvious thing to do in much of the country that one wonders why it has taken so long to be considered. In addition to water conservation, rainwater harvesting can help reduce stormwater runoff from building sites.

At the new Tacoma, WA, police facility designed by TCF Architecture, two 4,800-gallon tanks collect rainwater and recycle it for toilet flushing.

Imagine even a modest half-inch rainfall on a 24,000-square-foot roof. That event will generate 1,000 cubic feet, or about 7,500 gallons, of clean free water. In a climate like the Pacific Northwest, or anywhere that receives light rainfall a good part of the year, this system could be quite productive. Assuming one could collect 80% of annual rainfall of 35 inches, one would harvest about 420,000 gallons for reuse. Basic treatment with a sand filter and ultraviolet light would make it suitable for toilet flushing and similar non-potable uses. What could be simpler? Nothing, except that you can expect to pay $20,000 to $50,000 for such a system, and it's not in most building budgets.

But wait! That may not be the end of the story. Many urban areas have quite expensive charges for storm-drain hookups. I have found many cases where the impact fees or system development charges that are avoided by a 100% rainwater reclamation system were greater than the total cost of the rainwater collection and treatment system. In that case, a building owner is "money ahead" to install it. In one California university project, just the cost of installing the storm sewers to take water off the site and to connect to the town's storm drains was greater than installing two 20,000-gallon tanks to hold runoff from the *100-year rainfall* event and provide toilet flushing for a good part of the year.

One caution: don't expect rainwater to provide all of your needs, unless you are prepared to treat it to potable water standards and make that case to local code officials. In addition, the taller the building, the lower a percentage of annual needs the system will supply, because you've only got one roof for collection purposes, but more toilet and sink fixtures for each added story. Another caution: runoff from parking lots is often too polluted to collect and treat with simple systems, so you can't count on using it.

Rapidly Renewable Materials

Some people may be surprised that the use of rapidly renewable materials is included in the LEED rating system. But if you think about it for a minute, it's not so surprising. After all, why shouldn't we be looking for substitutes both for old-growth timber, tree-farmed wood and chemical substances such as vinyl composition tile (VCT). Each of these substances has some environmental issues associated with its production, harvest and use.

In the LEED system, the category of rapidly renewable materials in-

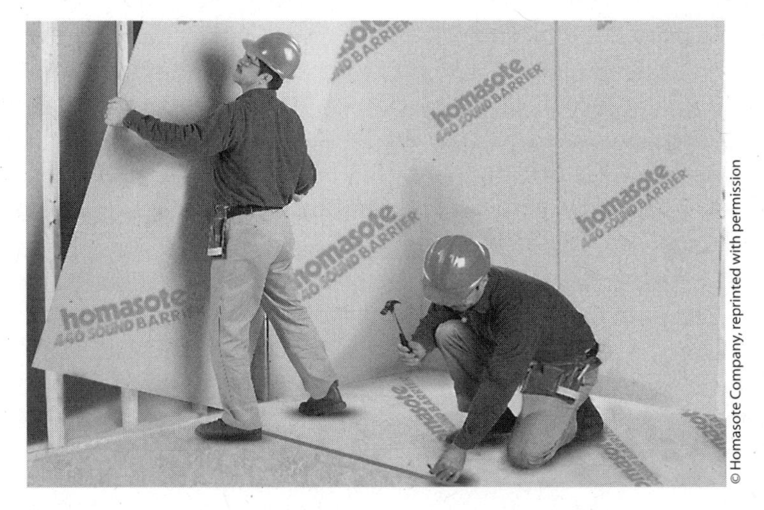

Homasote® panels from recycled newspaper installed in a building.

cludes anything that can be grown and harvested in less than ten years, such as agricultural panel boards from wheat, rice straw, sunflower seeds and sorghum stalks and used for cabinetry and wainscoting, interior doors, subflooring and even plywood; cork (cork typically has an ten-year regeneration cycle and comes from harvested tree bark in Spain and Portugal) and bamboo for flooring, linoleum floor covering and wool rugs.

Note that not all rapidly renewable materials meet other environmental criteria such as locally sourced (most bamboo flooring is from China, for example, and most linoleum from Europe), so it's sometimes necessary to make decisions about which values are more important to a project. Of all the rapidly renewable materials, bamboo seems to have penetrated farthest into the mainstream green building market. Growing up to 2 feet in a day and 60 to 80 feet in a year, bamboo can harden in 5 years to be equal in strength to a 50-year-old tree.[126]

Bamboo flooring is often chosen for its distinctive appearance and such desired qualities as hardness, resiliency and stability. Cork flooring is valuable for its sound and thermal absorption as well as its resilience. Linoleum is made from linseed oil, cork dust, natural fiber and wood powder — rapidly renewable and recycled ingredients. Project designers and product specifiers are critical components of the new green building "ecosystem," and they often have to consider issues other than the environmental attributes of a product, including performance, cleanability (especially for flooring) and durability. As they acquire more experience with new types of green products, many of these concerns gradually disappear.

To get a LEED point for using rapidly renewable materials, 2.5% of the total value of materials in a project need to be from this category. In a $10 million (construction cost) project, with $4.5 million devoted to cost of materials, this represents about $113,000 worth of such materials, about 1.1% of the total construction cost. These materials can be specified later in the design process, particularly where their aesthetics and provenance might prove valuable to the project's goals.

Recycled-content Materials

When you think of a green building, if you're like most people, recycling and building with recycled-content materials would likely spring to mind as a key characteristic. In the LEED system, credit is given to projects in which recycled-content represents more than 10% of the total value of all building materials (excluding equipment). This includes structure: roof, floors and load-bearing walls; rough and finish carpentry; insulation (sometimes cotton-batt insulation is made from recycled jeans); doors and windows; architectural metal, rebar, fly ash in concrete, steel structural beams, internal walls and floor coverings.

Because building materials typically represent about 45% of all construction cost, this standard implies, on a $10 million building, that $450,000 of recycled content would have to be used to qualify for one LEED point; for two points, $900,000 of recycled content materials would have to be used. Fortunately, some of the more expensive items, such as structural steel, have a decent recycled content (about 30%), while other steel items such as rebar (reinforcing rods in concrete) can have up to 90% recycled content.

The purpose of the LEED standard is to encourage the development of a local and regional economy that values recycling and that creates new materials with the same performance characteristics (strength, weight, durability, etc.) as virgin materials. In that way, we will begin to close the loop of resource flows in the economy and not incur the energy and pollution costs of using virgin materials.

An excellent example of a post-industrial or pre-consumer recycled-content material is the fly ash from coal-burning power plants. In concrete this material can be used to replace a significant amount (up to 50%) of cement, a very energy-intensive material to make, without compromising structural integrity. Cement makes up about 5% of concrete by weight, but it's the key ingredient. Another post-industrial material used to replace cement is the slag from blast furnaces.

Other examples of recycled-content products used in buildings include toilet partitions and exterior decking from recycled plastic bottles, Homasote® roof decking and wall panels from recycled newspapers (see p. 149),[127] acoustic ceiling tile, drywall with recycled paper facing, carpets made from recycled plastics, carpets with recycled fibers and ceramic tile from recycled glass.

● ● ●

Renewable Energy

There's so much to say about renewable energy. You can conceive it this way: imagine civilized life on Earth before the advent of the fossil-fuel era. Think of ancient Rome, the Italian Renaissance and the settlement of America, all of which took place using only sun, wind and water, along with draft animals, for power. Imagine that we could live healthy, happy and productive lives without electricity. This was reality for our great-grandparents (that is, if you're my age; if you're under 30, add another "great"). Think of Abe Lincoln in the White House, a once-habitable place to live and work without air conditioning (well, maybe not in July and August), gas furnaces and electric lighting.

As we start to bump up against the limits of planetary ecosystems to absorb all the waste and effluents made possible by fossil and nuclear fuels, we once again are reminded of the need to start living not off our inherited wealth, fossil fuels, but off our continuing income from the sun, wind, flowing water, geothermal energy and growing plants. Every 20 minutes, enough solar energy reaches the surface of the US to power the entire country for a whole year;[128] the problem is gathering it and using it economically.

For green buildings the most important renewable sources for onsite energy production are solar, wind, small hydroelectric (for rural uses near a river or stream) and geothermal (the Earth's heat). For powering our vehicles and cities, we can look to biomass power such as ethanol from corn production (and possibly plant residues) and, in the future, possibly solar-electric hybrid and electric cars. In the farther future, solar- and wind-generated electricity might be used for making hydrogen to power cars and buildings with fuel cells.

Counting wood burned for fuel and electricity, as well as large hydroelectric projects, wind farms and biofuels, renewables currently account for about 12% of total US energy use.[129] Many state legislatures are beginning to consider laws that would require electric utilities to produce 15% to 25% of their electricity from renewable sources within 20 years, by 2025 to 2030.[130] A newly formed group encourages government programs to help America's farms, forests and ranches generate 25% of the country's total energy needs from sun, wind and biofuels by 2025.[131]

From a design perspective, the key to using renewable energy economically in buildings is reducing overall energy demand through building orientation, passive solar design techniques, more efficient building envelopes (insulation and glazing) and more efficient equipment, both HVAC systems and lighting, as well as reducing internal "plug loads" such as computers, printers, copiers and refrigerators. That way, the renewable energy systems will supply a higher percentage of the building's total energy demand for the same cost than in a less-efficient building.

A 2007 report by the American Solar Energy Society showed that renewables such as solar, wind, biofuels, biomass and geothermal could supply a carbon reduction of more than 500 million metric tons per year, about 40% of the total needed to meet Kyoto treaty obligations.[132] Using onsite renewable energy, green buildings have a critical role to play in reducing our carbon footprint.

Renovation, Building

Why throw away an older building? There are many great examples of buildings that have been successfully renovated and restored to productive uses, even after standing dormant for many years. In Portland I witnessed the benefits of building renovation and reuse in a number of ways, since preserving older structures there was seen as a major civic and cultural value.

First, older buildings are typically part of the visual and cultural fabric of a neighborhood. By their very existence, they tell a story to everyone about earlier times. Second, when renovating, you are reusing materials

SERA Architects

For its headquarters in Portland, OR, SERA Architects renovated part of an older hotel building into a modern office in 2004 and received a LEED-CI (pilot) Gold certification.

that would otherwise require a lot of embedded energy to extract, harvest, process and transport to the building site. Third, there can be a significant cost to demolishing an older building, particularly in congested urban areas. Fourth, restoring older buildings often leads to further efforts in the same neighborhood, one way that cities revitalize themselves. Fifth, you are not adding building debris to landfills and causing more traffic in congested areas from waste pickup and disposal.

The LEED rating system recognizes the benefits of building reuse by awarding up to three points for maintaining 100% of a building's exterior and 50% of the interior structure, excluding non-structural roof materials and windows. It's often necessary to replace or upgrade a roof, and of course, you may want to install more energy-efficient windows in the building (replacing windows is harder for historic renovations). One drawback: older buildings were designed long before we knew much about earthquake engineering. So, in seismically active zones, there is often a significant cost associated with structural reinforcement to protect people and property in the case of an earthquake.

One final good reason: renovations can offer quicker occupancy, with less permit hassles, than tearing down a building and starting fresh. (This may not be true for historic renovations.) For homes, building renovation is part of the way in which cities renew themselves. Just as our bodies replace many of their cells at regular intervals (seven years is often cited), cities replace or renovate their housing stock at regular intervals, typically 50 years or so.

Restoration of Sites

A goal of many green building projects is to leave the land a better place than it was before. Achieving this goal requires site restoration activities, such as rehabilitating natural drainage systems, replacing wide swaths of green lawns with plants that provide wildlife habitat and replanting ornamental plants with native and adapted species that need far less water and intensive maintenance. As we complete the switch from a predominantly manufacturing economy to one based primarily on services, developers are finding attractive options in paved-over older parts of cities that once supported manufacturing, warehouses and similar industrial uses. Many of these sites were polluted with petroleum products, heavy metals, PCBs and other toxic substances that require remediation before reuse. Even paved-over but unpolluted sites can be converted to offices, retail, hospitality and housing, with considerably more wildlife habitat.

Often the task of the architect and builder is to find a way to place buildings so they don't disturb what's already working on a site. Several

GreenWorks, PC, Landscape Architecture & Environmental Design

Designed by Atelier Dreiseitl and GreenWorks, Tanner Springs Park provides wildlife habitat in an urban setting.

years ago, I visited the National Conservation Training Center of the US Fish and Wildlife Service in Shepardstown, West Virginia. Located on an upper tributary of the Potomac River, this site is very hilly, like most of the state. The project designers placed 17 buildings on the site, only on the hilltops, leaving the hollows alone. Because of a number of changes of site elevation, the design required many wooden bridges between buildings, sometimes with entrances on upper floors. However, this approach allowed the project to avoid extensive grading and degradation of wildlife habitat, while promoting the very values inherent in the Center's mission. This is a good example of a smart and wise approach to site planning.

Another interesting project is Tanner Springs Park in Portland, Oregon. Completed in 2005, this park sits on top of about 40 feet of historic fill of the original Tanner Creek. To honor its origins and to provide city residents with a natural park, the landscape architects designed a reconstructed wetlands with a boardwalk over it. This park is now habitat for many creatures including various waterfowl. It is mainly used for passive recreation and helps incorporate sustainability into the fabric of the city.

Return on Investment

Green buildings eventually will compete in the marketplace with standard buildings, so it's reasonable to ask that they be evaluated financially and economically on the same basis. Return on investment (also expressed as internal rate of return or net present value) is fundamental to evaluating economic decisions. Simply put, return on investment is how much I plan to make, either annually or totally, from an investment, with all numbers expressed in today's dollars. If I'm going to take the risk of making an investment, then my goal should be to equal the return from similar investments.

Saving energy is like buying a long-term bond. If I invest $100,000 today in a ten-year treasury bond, I expect to make about 5% each year, before taxes and inflation, a return dictated by the bond market. This is a completely risk-free investment, but my net gain is small (imagine 20% combined state and federal taxes and 3% inflation, leaving me with a net real return of 1%).

If I invest $100,000 in energy-savings improvements that yield $20,000 per year in savings (a five-year payback on the incremental investment), my return is 20% per year, almost risk-free (as long as the building continues to be occupied and energy costs don't fall), and my return goes up typically at the rate of inflation or better, so I get to keep more of my gain than with a bond.

Green buildings offer other returns than just energy savings. If they have a lower vacancy rate, command higher rents or sales prices, higher renewal rate at the end of the lease, water savings, lower insurance costs, etc., then these virtues have an economic value to an owner or developer. Considering also the tax credits and deductions for investing in energy-saving and renewable energy technologies — which may exceed $1 to $2 per square foot on a new building — there is an excellent economic case for private owners to invest in green buildings, based strictly on return-on-investment criteria.

In my book, *Developing Green: Strategies for Success*, I calculate the benefits of such green building benefits.[133] For example, a lower vacancy rate of just 2% translates into a building value increase of 3%. In the world of commercial real estate, this is not a trivial gain.[134]

David Gottfried, one of the co-founders of the US Green Building Council and a former Washington, DC, real estate developer, gives an inspiring talk about return on investment for green real estate projects. He talks not about the bottom line but about the top line, the positive effect on project revenues of having a more desirable project in the marketplace that enhances the ability of a project to attract high-quality tenants, get better rents and keep the tenants for longer periods of time. Also, there is a strong return on investment for an organization committed to sustainability as a basic value proposition, in terms of recruiting and retaining high-quality employees and business associates. What makes you get out of bed in the morning is typically not your salary, but the purpose of the organization and the value or importance of your work, isn't it?

Right-sizing Systems

I call this the "Goldilocks principle" of building design: systems should be the right size for the job, neither too big nor too small. This seems obvious, but since temperature complaints are a large source of occupant discomfort and criticism in most buildings, engineers have become naturally conservative in sizing systems, making them larger than they need to be, both to avoid complaints and to provide an extra cushion for extreme events and the degradation of system performance over time. This decision costs a lot of money, since HVAC systems alone can cost 10% or more of a building's budget.

Right-sizing a system doesn't start with the HVAC system. It begins with proper building orientation; generally in the US and Canada, buildings should be oriented with the long axis in the east-west direction, allowing for less heat gain through the east and west windows, when sun angles are lower. This measure alone can save 10% or more of the energy use of a similar building oriented in the north-south direction. For better daylighting, buildings need to be more rectangular than square, But unless daylighting can be sold as an important feature of an office building, this is not usually considered an "efficient" design from a developer's perspective, because it results in a smaller percentage of leasable area compared with a square building design, so there may be a built-in conflict here between economic efficiency and good design practice.

Then one needs to consider how to keep direct sunlight out of the building in summer and to let it in during the winter, a basic tenet of passive solar design. This can be done with selection of glazing that restricts incoming sunlight or with external shading devices that keep the high summer sun from entering the windows. Thermal-energy storage systems are used to reduce the demand for cooling during summer afternoons, often a leading cause for over sizing systems. In large projects, it may be practical to include space for installing additional cooling capacity if the type of occupancy changes, just to be on the safe side, rather than spending the extra money upfront.

Andy Frichtl, a senior engineer at a mid-sized consulting engineering firm, Interface Engineering, Inc., says that right-sizing requires engineers to go back to basics and develop system sizing from careful analysis of heating and cooling requirements rather than relying on handbooks and rules of thumb (such as 300 square feet of building area per ton of air conditioning) that already have too many safety factors hidden in them, justifying systems larger than necessary.[135]

Why should engineers do this? In a sustainable design project, each contributor needs to be mindful of the need to save money wherever possible to fund those green measures that are definitely going to cost more, such as photovoltaics for onsite power production or a green roof for open space, stormwater management and habitat restoration. In addition, buildings designed this way will be cheaper to operate, use less energy over time and may well be more comfortable for the occupants, leading to higher productivity and greater satisfaction.

Salvage Materials

"A penny saved is a penny earned," wrote Ben Franklin. We care about reusing building materials because of the energy and resources they represent. It takes energy to down-cycle them into recycled-content materials (think of old concrete from a building ground into three-quarter-inch aggregate for use in concrete or as the base material for a parking lot or roadway), so why not use them in their original form instead of throwing them away or using them in some devalued form?

LEED recognizes the value of salvaged or reclaimed materials, such as decorative brick, heavy timbers and other framing lumber, doors, millwork, furniture and partitions, by rewarding projects that use them for at least 5% of the total value of all building materials (not counting equipment). On a typical $10 million (construction cost) project, this would represent $225,000 worth of such materials, not an insignificant amount. One benefit of this practice is the development of local enterprises based on deconstructing buildings and salvaging such materials. If you consider how much useful material is saved from old cars by auto salvage yards in every town, you'll see the benefit of this practice.

With the advent of Web-based auction sites such as eBay and retail/wholesale reclaimed building materials stores in most large metropolitan

Shane Shukis, UCR ARTSblock, University of California-Riverside

At the University of California, Riverside, an older building was deconstructed and most of the materials salvaged by artist Jason Middlebrook for making furniture and other useful products.

areas, there is now a nationwide market in reclaimed building materials for building projects. So, there is no longer an excuse for not being able to find materials. The only issue is their quality and availability, along with transportation and storage costs.

Some creativity might be required to find and reclaim salvaged materials. The first LEED Platinum project, the Chesapeake Bay Foundation building in Annapolis, Maryland, used large wooden tanks from a former pickle manufacturing facility to harvest rainwater from the roof of their new building. The three tall pickle barrels create a strong visual and architectural element at the building entrance.

●●●

Schools, Green

In the late 1990s, researchers in California conducted a classic study of the benefits of daylighting. Analyzing the test scores of more than 21,000 school children in the Seattle area, Fort Collins, Colorado, area and southern California and correlating the scores with the amount of window area, daylighting and views to the outdoors, their study showed conclusively that test scores went up 7% to 21% when kids had more windows, daylighting and views outside.[136] This study put to rest probably the worst idea in school design, the advent of windowless classrooms in the 1970s to save energy and keep kids focused on their boring lessons. A follow-up study of 8,000 children in 450 in elementary classrooms in central California showed similar results. However, classroom design was critical to obtaining these results: daylighting and windows need to have measures to reduce glare and direct sun penetration, to have photosensor controls for daylighting and to have good thermal comfort controls, especially in a warmer climate.

Many people greeted this study with incredulity that such a simple measure could have such dramatic results. I'll lay odds that the majority of so-called educators, school superintendents and school boards still don't know it exists. Beyond the effect on kids' health and performance, you'd probably find that more daylighting and views to the outdoors is also healthier for teachers and other staff.

I've wondered why this result should be surprising. The last thing most kids want is to be cooped up for hours each day in poorly lit classrooms with few windows. Why do we consign our most vulnerable population, on which we Boomers are depending for Social Security benefits and support in our old age, to such poor conditions? I'd like to think the reason is ignorance, not incompetence on the part of architects and school administrators. They've got a lot on their plate, after all, just to get schools built. But I'm not so sure daylighting is a high enough priority for most architects.

My friend Heinz Rudolf is a well-seasoned and very talented architect in Portland. He's been designing beautiful schools with abundant daylighting for at least the past decade. He believes so strongly in the necessity of daylighting that he will fight with his clients to get it included in each school. The figure below shows one of his projects in the suburban town of Clackamas, Oregon. His designs are energy-conserving as well, and — guess what? — he gets them built at the average cost of similar schools in the area. If green schools don't cost more to design and build, why isn't everyone creating them as a matter of professional best practice?

Earlier this year, I got a call from a self-described "activist mom" in Pennsylvania. Her local school board was getting ready to build a new middle school and wanted to recycle a set of plans from the last school built, in 1990, to save a pittance on architectural fees (about 1% of a $35

Michael Mathers

Clackamas High School, Oregon, designed by BOORA Architects, featuring abundant daylight. Certified at the LEED-NC Silver level in 2002, this project is one of the first LEED-certified schools in the country.

Financial Benefits of Green Schools ($/sq.ft., 20-year net present value)

Energy cost reductions	$9.00 (30% average savings)
Emissions reductions	$1.00 (from reduced energy use)
Water and wastewater costs	$1.00 (32% average water savings)
Asthma reduction	$3.00 (25% reduced incidence)
Cold and flu reduction	$5.00 ($45 per child per year)
Teacher retention	$4.00 (3% reduction in teacher turnover)
Employment impact	$2.00 (from higher initial costs of green)
Future increased earnings	$49.00 (from better health, higher test scores)
Total benefits	$74.00
Cost increase	($3.00) (assumes 2% higher initial cost)
Net benefit	$71.00

Capital E

million building project that will last 50 to 75 years). Imagine the lack of daylighting and energy efficiency in a nearly 20-year-old set of building plans, and you (like her) will be astonished at the lack of priority placed on the health of the children when it comes to school building design.

A 2006 research report by Greg Kats on the benefits of green schools should put to rest the notion that green design is an option that we should use only if we have extra money for a project. Given the clear and compelling evidence, doesn't the act of *not* building green schools, at a minimum with abundant daylighting and views to outdoors, represent professional neglect?

But there's even more to the story. The table above shows the range of benefits you can expect from green schools.[137] In addition to the operating cost benefits you might expect from energy-efficient schools, look at all the others you might not have considered at first glance. There is a total benefit of about *25 times* the assumed initial cost increase. Throw out even the imputed benefit of increased lifetime earnings for kids who have greater learning, more time in the classroom and higher test scores. There is still $25 of benefit for a $3 investment, about *8 times* the return over 20 years. So who is being unintelligent: those who advocate for spending the money to build green schools or those who oppose the idea because it supposedly costs more? Even the $10 per square foot of benefits that accrue directly to the school (from energy and water savings) are triple the assumed higher initial cost.

S

Solar Thermal Systems

I got acquainted with solar energy systems in the mid-1970s when the first Arab oil embargo led many people to start looking to the sun as a source of free energy (as the slogan went, "Four billion years without a shortage."). I participated in the development of the solar energy industry in California over the next decade, first directing the state's landmark solar-industry commercialization programs, then as a private sector participant, marketing, selling and installing solar systems for water, space and pool heating. The current upsurge of interest in solar energy systems is the first in 20 years. The domestic solar thermal industry collapsed in the mid-1980s, the victim of falling oil prices, low natural gas prices and the expiration of federal and state tax credits first enacted in the 1970s.

During the period 1975-1985, solar thermal technologies were thoroughly explored. In California I estimate that more than 250,000 solar water heating systems were installed for homes, apartments and factories, along with tens of thousands of pool heating systems (one part of the industry that never collapsed) and thousands of solar home heating systems. In almost all cases, conventional heating systems were used as backups during winter and periods of low sunshine and cloudiness. In fact solar thermal technology has been used in some form for more than 100 years in the US and elsewhere.[138]

The federal Energy Policy Act of 2005 gave new impetus to the solar thermal industry, providing (currently through the end of 2008) federal tax credits of 30% for residential and commercial solar water heating and space heating systems. In most parts of the US, solar water heating systems can easily supply 50% or more of annual requirements for a family home, apartment house or business.

During this past decade of growth in green building, I have wondered why more explicitly sustainable projects don't use solar thermal systems. They work quite well, there are plenty of local suppliers, the economic benefits are reasonable (with federal tax credits, along with many state incentives, you get typically a 10% or better annual return on investment), and they represent a visible symbol of commitment to renewables that most people can readily identify. Perhaps the reason is that architects and engineers doing commercial work are simply unfamiliar with them and reluctant to experiment.

From the standpoint of basic physics, using a very low-intensity energy source such as solar is intuitively appealing. With a single-glazed, metal-finned collector, the sun can easily heat water to 160°F, more than enough for a typical water heater (usually set about 125°F to 135°F), and protecting the collector against freezing is not difficult in most climates. Many of you may have used an elevated black-plastic water bag as a way to have a hot shower while camping. The technology is that simple. Many parts of the world that have to rely on diesel or heating oil (or even wood) for water heating have used solar water heating for decades, including a good part of the Mediterranean and Australia. On a trip to Greece a couple of years ago, I saw many solar water heaters atop apartment buildings in Athens, a testament to the basic utility of this technology.

● ● ●

Stormwater Management

A basic principle of sustainable design is that buildings should be able to supply all their water needs from the annual rainfall on the project site and from recycling the wastewater generated by a project, effectively getting several uses out of the same amount of rainwater. At the present time, it's not always possible for projects to use all of the rainfall on a site, so they need to reduce the impact of new development on downstream flows from stormwater running off a site.

The problems with pollution of lakes, rivers and the ocean from urban runoff were mentioned previously. In addition, new development can tax existing stormwater collection systems with increased runoff and higher peak flows from paved surfaces. On a large scale, urban development has led to greater flooding, both from higher runoff and from more people locating in flood plains.

As a consequence, the LEED system deals explicitly with stormwater management by rewarding projects that reduce both the rate and quantity of stormwater from a site and those projects that improve the quality of runoff generated from site runoff. LEED rewards projects on previously undeveloped sites that keep runoff from the most frequent storms (those that occur, on average, every two years or more frequently) to pre-development conditions, by instituting various measures such as detention or retention ponds, onsite infiltration (using permeable paving), bioswales (vegetated runoff ditches), green roofs and using native and adapted vegetation instead of turf.

On sites that are already more than 50% impervious (for building

Bioswales can be integrated with other architectural design features to take rainwater from buildings, either to remove from the site or to recycle for building water uses, as in this project at Portland State University's Epler Hall, a LEED-NC Silver-certified building, designed by Mithun architects.

additions or renovations in most developed urban areas), LEED rewards those that reduce stormwater runoff by 25% or more from prevailing conditions. To this end, green roofs are a great aid, especially on limited urban sites that usually don't have extensive areas for new plantings.

Sustainable Design

According to author and architect Jason McLennan, "sustainable design is a design philosophy that seeks to maximize the quality of the built environment, while minimizing or eliminating negative impact to the natural environment."[139] This is a succinct and valuable definition, one that complements the oft-quoted statement of the UN's Brundtland World Commission on the Environment and Development in 1987, "Our Common Future," defining sustainable development as meeting "the needs of the present without compromising the ability of future generations to meet their own needs." Sustainable design is also a major movement in contemporary architecture and engineering practice. Above all, it is about "the way things are used; how they are communicated to the world; and the way they are produced."[140]

Elements of sustainable design practice include:

- High levels of resource efficiency overall, including transportation and energy use in building materials, construction and building operations.
- Energy-efficient building systems.
- Renewable energy use.
- Water conservation and graywater reuse.
- Habitat preservation and restoration.
- Use of natural energies for building heating and cooling.
- Rainwater capture, reuse and recycling.
- Natural stormwater management.
- Use of recycled-content, non-toxic, salvaged and local materials.
- Healthy and productive indoor environments for people.
- Durability of building materials and designs.
- Flexibility for building uses to change over time.
- Access to alternative transit modes.

Sustainable design considers the big picture: the need to transform global settlement and industrial patterns to be healthier and less wasteful, less impactful on the natural environment. It brings these concerns down to the scale of each building, each site plan, each choice of materials and processes. When successful, sustainable design is often hard to detect. A building just feels right on the site rather than obtrusive; there is abundant daylighting; nature is both within and outside; as a design element, water flows naturally from the building into a bioswale or other natural drainage feature; the building is comfortable without a huge rush of moving air; internal spaces create expansiveness and delight; and the overall effect is beautiful. As many scientists have noted, if the solution to a problem is not "elegant," it is either incorrect or there is probably a simpler solution waiting to be found.

Sustainable Sites

If it's on a poorly selected site, can it still be a green project? In general the answer is yes, but green building assessment systems such as LEED give guidance for site selection. The goal is to develop only on appropriate sites and to avoid the environmental impacts of locating on poorly chosen sites. Green building projects should avoid locating buildings, hardscape (paved surfaces for landscaping such as plazas and walkways), roads or parking areas on sites that:

- Are previously undeveloped and less than five feet below the 100-year flood level (this requirement can be met in portions of older cities situated alongside rivers or lakes that are already developed and that are in the 100-year flood plain; the LEED requirement covers only previously undeveloped sites).
- Are located on prime farmland (this point would not be available for a lot of suburban development that is gobbling up prime farmland all over the country).
- Are within 100 feet of a wetland, as defined by state and federal regulations, or at a greater distance, as provided for under local laws or zoning ordinances, even if previously developed.
- Are designated as habitat for any threatened or endangered species (this is generally prohibited anyway by state and federal laws).
- Are located on previously undeveloped land that is within 50 feet of a water body, defined as seas, lakes, rivers, streams and tributaries that support or could support fish, recreation or industrial use (the idea here is to keep buffer zones around water bodies so that the public can access them, and so that development has lesser impacts by being set back).
- Are situated on land that prior to acquisition for the project was public parkland, unless land of equal value as parkland is accepted in trade by the public landowner (local, state or federal park projects are still allowed to build, for example, visitor centers to green building standards).

In addition to site selection rules, LEED also encourages projects to locate in areas of higher density with at least two-story development (60,000 square feet per acre or more) in the surrounding area or to locate in areas that have a number of important services available to residents of apartments or occupants of offices and shops. In this way, the infrastructure will already be in place, and with enough services accessible within a walkable distance (half a mile or less), people will be encouraged not to drive to get to such essential services as a drug store, convenience store, bank, dry cleaner, park, doctor's offices, schools, restaurants, department store or mall, child care and health clubs or fitness centers.

I worked in downtown Portland for nine years and found it absolutely delightful to take care of life's many small errands on my way to work, during a coffee break, during the lunch hour and right after work. When I drove to work, which was not often, my car would often stay parked for the entire day, and I could still get everything done on foot. It was a much healthier way to live than working in the suburbs and having to drive everywhere.

t

Technology, Green

During 2006 and 2007 venture capitalists finally discovered green building, investing in a wave of product innovations and renewable energy, which attracted considerable money in 2005 and 2006 when oil price hikes reached record levels. In this $1.1 trillion building industry (construction value), private-equity companies are investing in products and systems that reduce the use of toxic materials in construction; make the construction process more efficient; encourage green housing technologies; provide for environmentally friendly methods for salvaging timber; increase energy efficiency in building systems; provide more ways to insulate homes and buildings with concrete, for example, by using insulated concrete forms (ICFs); and use energy-efficient structural insulated panels (SIPs).[141] There is also growing interest in using nanotechnology in such areas as photovoltaic cells, indoor thermal-insulation coatings, self-cleaning glazing and stronger steel for rebar in concrete construction.[142]

We're seeing the world's largest companies make major commitments to energy technology and a wave of other innovations that will affect building design. A good example of this is the $1.5 billion investment *ecomagination* campaign launched by General Electric's (GE) CEO Jeffrey Immelt in 2005, well before the current wave of eco-inspired corporate concern. The campaign includes major investments in solar and wind energy technology as well as in water desalination. In addition GE set a 30% greenhouse gas (GHG) intensity reduction goal by the end of 2008, along with a 1% absolute reduction by the end of 2012. GE has also set an energy-efficiency improvement goal of 30% by the end of 2012. Progress will be measured against a 2004 baseline.[143]

The growth of exhibit booths at the US Green Building Council's annual Greenbuild trade show and conference also highlights the wave of technological changes sweeping the green building industry. From a modest beginning of 220 exhibit booths in 2002, the 2006 show featured 700 exhibit booths, more than a threefold increase in just four years.[144] The November 2007 show, to be held in Chicago, expects to feature 850 exhibit booths and to attract nearly 20,000 attendees.

In the sphere of building design, many new technologies are appearing that allow architects and engineers to specify far more energy-efficient products, for many more uses, and to analyze the impact of these measures on a project's energy efficiency much earlier in the design process. We're also seeing 3-D modeling become a reality, so that new passive design techniques can be examined for their energy impacts before designs are hardened into working drawings. In the area of climate control, just as underfloor air distribution systems came into use in the late 1990s, we're now seeing a variety of other building climate management systems, including "double-envelope" renovations of older buildings in cold-climate regions to benefit from natural ventilation in cold weather. A double-skin façade at a new research center on the University of Toronto's St. George campus is an example of this trend.[145]

Thermal Energy Storage

Thermal energy storage is a simple concept: make ice or chilled water when power is cheap, then avoid buying electricity to operate mechanical cooling systems when it is expensive. Just about any large office building, hospital, hotel and similar 24/7 facilities can benefit from thermal energy storage. As utilities are forced to increase peak-period electrical rates to

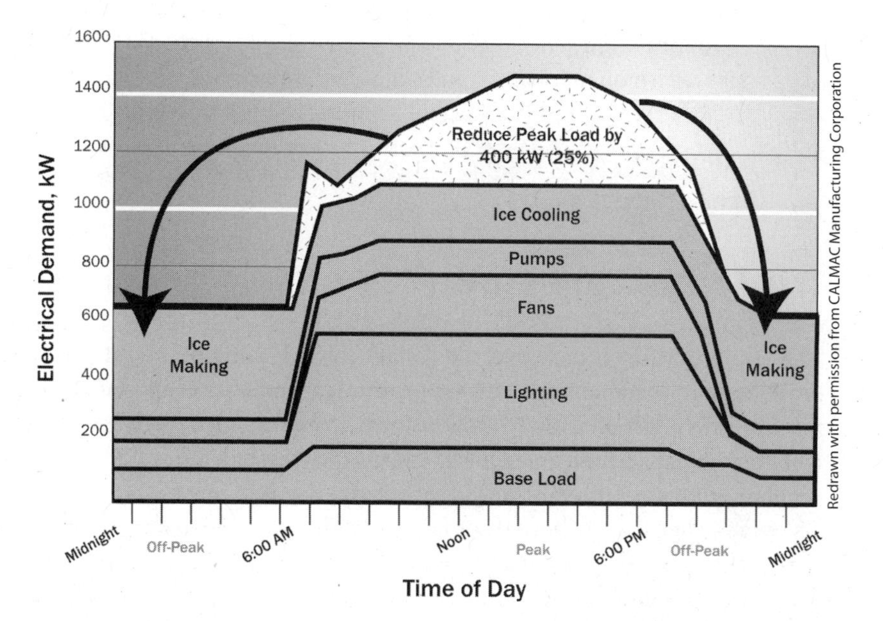

Thermal energy storage systems can shave peak power demand by 25% in large buildings and facilities.

avoid brownouts and blackouts on summer afternoons by limiting demand, thermal energy storage systems are beginning to come into their own. They reduce the pressure on the electrical grid and can save building owners considerable money. They also can save money upfront, even after paying for their cost, by limiting the size of the HVAC system purchased. They should be thought of as a component of an integrated design project, where the initial focus is on reducing the need for summer air conditioning through overhangs and shading devices, better glazing and passive solar design approaches, not as an end in themselves.

Consider the example shown in the table on page 169. By applying thermal energy storage, the building's electrical demand is lowered by 400 kilowatts, about a 25% reduction in peak demand. If the utility charges $10 per kilowatt, not an unusual amount, the monthly savings from demand avoidance alone could be $4,000, or nearly $50,000 per year! These savings can continue for many months beyond the peak summer cooling period, because utilities often charge for the instantaneous peak power use in a quarter, or sometimes in an entire year. If time-of-day rates are available from the local electric utility, making power cheaper to buy during off-peak hours, the energy cost to produce the cooling would also be lower, further increasing the cost savings from installing the system (only about 10% of ice storage systems provide the full cooling load). In themselves, electric utility "demand avoidance" charges provide complete economic justification for partial storage systems. Time-of-day rates are not needed, but a faster return on investment will occur with them.

● ● ●

Triple Bottom Line

The triple bottom line represents a new way of thinking about green buildings and sustainability: truly appropriate measures need to incorporate consideration of ecology/environment, economy and (social) equity (the three Es), or people, planet and profits (the three Ps). The triple bottom line contrasts with the traditional business bottom line which considers only profits. Clearly businesses have been moving away from a profits-only orientation for quite some time, as the corporate social responsibility movement has taken hold: getting and keeping good employees who care about the environment and about the local community has become a central element in profitability.

In the past few years, the green building movement has pushed many companies to rethink their entire approach to development, emphasizing energy efficiency and LEED ratings in their real estate offerings. Good examples of this movement abound. At the 2007 "Green Cities" conference in Sydney, Australia, the entire large property development industry in Australia appeared lined up behind the idea that green buildings promoted higher property values and enabled them to acquire better tenants and buyers. Late in 2006 prominent real estate consultant Charles Lockwood claimed that "trillions of dollars of commercial property...around the world will become obsolete — and will drop in value" because green buildings are going mainstream.[146]

My book, *Developing Green: Strategies for Success*, presents the overwhelming business-case benefits of green buildings.[147] These benefits include not only the expected utility savings, but also improved health and well-being of employees, easier property marketing and better public relations, ease in getting both debt and equity capital for development, and improved retention of key employees. There is no better example of how the triple bottom line works than green buildings. We can improve the urban and natural environment, save energy, reduce infrastructure investment costs and gain better health and morale of the workforce through the simple act of better building design. So, what is stopping us?

Architect William McDonough also talks about the "triple top line": thinking of these three things together allows one to grow profits while also improving environment and enhancing equity. In other words each aspect wins, and there is no tradeoff required between economic efficiency, environment and social equity. If you find this is not the case, you have to go back to the drawing board and try harder! In a 2002 essay McDonough and his collaborator, Dr. Michael Braungart, state:

> A business strategy focused solely on the bottom line, however, can obscure opportunities to pursue innovation and create value in the design process. New tools for sustainable design can refocus product development from a process aimed at limiting end-of-pipe liabilities to one geared to creating safe, quality products right from the start.[148]

U

US Green Building Council

The US Green Building Council (USGBC) was formed in 1993 by a few enterprising souls who wanted to transform the building marketplace into a more environmentally responsible activity.[149] Consisting solely of organizations, the USGBC now (early 2007) represents more than 8,500 members including federal, state and local government agencies; colleges and universities; environmental NGOs; product manufacturers; trade associations; architects, engineers and builders; and a myriad of other disciplines and professions engaged in the building industry. With less than 100 paid staff people, the USGBC is heavily supported by tens of thousands of hours annually from volunteers representing member companies who staff the extensive committee system that guides the organization's technical efforts. There is also a widespread system of more than 70 local chapters. The Canada Green Building Council is an equally dynamic force for change; in per-capita terms, its early 2007 membership of nearly 1,400 exceeds that of the USGBC.[150]

As its earliest priority, the USGBC developed the LEED rating system to define what made up a green building. Seven years after the introduction of the LEED version 2.0 in 2000, more than 5,000 projects are now registered under the LEED system. At an average building size of more than 100,000 square feet, LEED projects represent about 500 million square feet of construction, equivalent to about 20% of the annual commercial square footage constructed in the US.[151]

In addition to certifying a building's greenness with the LEED rating system, the USGBC trains people in using the scheme. As of early 2007 nearly 45,000 had been trained, and more than 35,000 had passed a national exam to become LEED Accredited Professionals. It is amazing how many people, from all different aspects of the building industry, are interested enough in this rating system to attend an all-day workshop and to take a test to certify their capacity to work within it. Through the LEED rating system and the training of building professionals, USGBC is effectively building the capacity for a major shift in building design, construction and operations, something that would have been inconceivable at the beginning of the decade.

In 2007 the organization responded to the climate change challenge by changing the LEED rating system to require certain minimum levels of energy efficiency from all certified projects. Beginning in 2007, as a result of these changes, USGBC leadership expects that LEED-certified buildings will reduce overall carbon dioxide emissions by 50% compared with conventional buildings.

USGBC believes we have reached a tipping point in the green building revolution. The organization's CEO, Rick Fedrizzi, predicted in late 2006 that, by the end of 2010, there will be 100,000 LEED-certified buildings and one million LEED-certified homes, a major increase in such projects in just four years.[152] It's clear that the USGBC is one of the major catalysts for green buildings in the US and that its membership and influence will continue to grow significantly in the next few years. (Shouldn't your university, government agency, company or non-profit become a member?)

● ● ●

Unbridled Enthusiasm!

This may seem like a strange term to include in a book about green buildings, but it is important nonetheless. I have never met a more enthusiastic group of professionals than the green building crowd. Enthusiasm is a key quality in making any major changes in existing systems. Stemming from the Greek phrase *en Theos*, "possessed by a god" or "God in me," enthusiasm helps overcome all obstacles.

The 19th century American philosopher Ralph Waldo Emerson said, "Nothing great was ever achieved without enthusiasm." For Emerson, enthusiasm gave a person courage to follow her inner direction. "Whatever course you decide upon, there is always someone to tell you that you are wrong. There are always difficulties arising which tempt you to believe that your critics are right. To map out a course of action and follow it to an end requires courage."

David Gottfried was a successful young property developer in Washington, DC, when he got the urge to found the US Green Building Council in 1993. He later founded the World Green Building Council in 1999, which now numbers ten member countries, US, Canada, Australia, New Zealand, India, United Kingdom, Mexico, Taiwan, the United Arab Emirates and Japan. David's enthusiasm for green buildings is infectious; as a result of his leadership acumen, persuasive skills and incredible drive, the USGBC is now the most dramatic force for change that the building industry has seen in decades.

A Canadian, Joe van Belleghem is an accountant by training. Currently, he is developing Dockside Green, promised to be the world's largest all-LEED Platinum development in Victoria, British Columbia.[153] Anyone who listens to Joe talk about how doing the right thing is the best business proposition comes away convinced that they need to do the same. To win the competition to build Dockside Green, Joe's development team beat out a better-funded team, in terms of resources and reputation. Joe's enthusiasm won the competition and continues to overcome all obstacles to the project's success.

Kath Williams holds a doctorate in education and was director of research at Montana State University when she got the green building bug well into her professional career. Working with national green building leaders, she sponsored one of the pre-eminent green building projects of the 1990s, the EpiCenter at Montana State. Now an international green building consultant heavily involved in the World Green Building Council and past vice-chair of the USGBC, Kath is a sophisticated, no-nonsense cowgirl from Montana who can convince anyone that not building green is incredibly stupid.

Meet Bob Berkebile, a founding partner of a leading architectural firm in Kansas City, Missouri, and a co-founder of the AIA Committee on the Environment in 1989, whose skillful rendering of sustainable design into a number of landmark projects continues to inspire new generations of architects and engineers. Bob seems to gain more enthusiasm for green buildings with each passing year. If the green building movement has the equivalent of *Star Wars*' Yoda, the wise elder, Bob is it.

Urban Heat-island Effect

The urban heat-island effect was first documented in the 1960s when scientists noticed that cities were noticeably hotter than the surrounding countryside. Sometime during the winter, during your morning commute, listen to the weather reports, and you'll see that the urban areas are typically 5°F to 7°F warmer than the more outlying areas. This is pretty natural, if you think about it. We pump huge amounts of energy into cities, and as every high-school physics student knows, all energy eventually becomes waste heat. In addition we pave over our cities, remove vegetation and put up concrete structures that retain the sunlight well into the night. We could not have designed a more perfect way to heat up the cities.

Partly as a result, about $40 billion is spent every year by property managers and building owners to air condition buildings.[154]

In winter this might sound good, just as global warming might sound good to some people in Alaska (until they consider all the consequences). But in summer it creates a much hotter local microclimate in cities, with less cooling from transpiration by vegetation and more air conditioning (which cools buildings but puts all the heat into the street or the air). In Phoenix, reportedly, by the year 2000 nighttime low temperatures in summer were 10°F hotter than in the 1970s. It other words, it never cools off for months on end. This means electricity demand is much higher in summer, which in turn drives the surging demand for new power plants.

What can be done? LEED suggests several simple measures to reduce solar heating in the summer and create cooler microclimates. First, reduce the amount of hardscape or impervious surface areas that can absorb heat. Second, put shading around all absorptive hardscape surfaces (parking lots, sidewalks, patios and plazas) so that at least 50% is shaded at noon on a typical summer day. In some areas such as the South, vegetation around parking lots can get pretty tall within five years! Third, place 50% or more of parking underground or under the building so there is less surface area to heat up from parking lots. Fourth, use highly reflective paving materials (gray or white concrete instead of asphalt) for a parking lot so that more incoming solar radiation will be reflected back into space and not be absorbed. Fifth, use an open-grid pavement system that would have vegetation growing inside the pavers so that there is less area to absorb heat from the sun. Obviously these measures can be used in combination: you can see there are many options for reducing the urban heat-island effect from hard surfaces.

When it comes to the building itself, there are two major options. Elsewhere we have profiled cool roofs and green roofs. LEED requires a green roof to cover at least 50% of the roof surface and a cool roof to cover at least 75%, or some combination of the two approaches. Either will reduce not only the urban heat-island effect, but also a building's demand for air conditioning in summer by providing cooler attic or subroof temperatures.

Vastu Shastra

Similar to Feng Shui, a design philosophy and approach known as Vastu Shastra is based on teachings of the ancient scriptures of India. Introduced into the US by the American practitioners of Transcendental Meditation (TM), Vastu Shastra aims to harmonize people, buildings and land. While individual homes have been built using these ancient principles, the largest commercial expression of this approach to creating healthy work and living environments is an office building in Rockville, Maryland, expected to open for business in 2007.

Located on 11 acres, the $72 million, 200,000-square-foot 2000 Tower Oaks Boulevard building is aiming for a LEED Gold rating and an Energy Star certification, based on a projected 41% savings in energy use. The building will also be the headquarters for two privately owned real estate companies, The Tower Companies and Lerner Enterprises, who will together occupy about 70,000 square feet.[155]

Vastu Shastra focuses on building orientation, facing east to take advantage of the energy of the rising sun. According to Vastu Shastra philosophy, this helps to energize people as they come to work. Many Native American dwellings also opened to the east to capture the warmth and energy of the rising sun. (By contrast, most passive solar designs have the long axis of a building facing east-west, and not north-south as is the case here, to promote energy efficiency and daylighting.)

Different activities are located in the building to take best advantage of the sun during the day. This approach is easier to apply to homes, where kitchen, bath, living room and bedroom all are used at fairly specific times of the day; in a commercial building, most uses don't really change much during the daylight hours.

Vastu Shastra is also concerned with proportion, a key to successful design in nature. Proportion is a key principle in architecture, and nature often mimics the same proportions at different scales, an approach characterized by fractal geometry. As you might suspect in a healthy building, Vastu Shastra also has a focus on using natural and non-toxic materials, filling rooms with daylight and fresh air and using the sun's power for onsite energy generation. As befits a healthy building, 2000 Tower Oaks Boulevard will also have an onsite fitness center, a meditation room (by the way, you can find these in a number of airports now, so why not have them in office buildings?) and a nature preserve, to bring nature closer to people and to preserve what is already there. What solar doesn't provide, wind power will; the project plans to fulfill 100% of its power needs from renewable energy sources.[156]

You can see that Western architecture can certainly learn a lot from traditional Eastern design methods, with their focus on managing subtle, as well as physical, energies and on harmonizing dwellings and buildings with the natural environment. Living and working in healthy, sustainable environments has long been a focus of other cultures, one we are just beginning to emulate.

● ● ●

Ventilation

Everyone knows the importance of fresh air. Just stay cooped up for a few hours in a poorly ventilated room and see how sleepy and generally out of sorts you feel. Engineers design buildings to meet or exceed building code requirements for fresh air. LEED gives an extra point for designs that circulate at least 30% more fresh air than standard approaches to encourage this simple approach to a healthier building.

In many parts of the US, fresh air can also cool buildings for hundreds of hours a year, without using air conditioning. A well-designed and energy-efficient building will make maximum use of the free cooling of outdoor air, so long as humidity is not excessive or heat loss too great (remember that for every cubic foot of fresh air coming into a building, a cubic foot of conditioned air has to leave).

But there's more to the story. Many buildings in windy areas can take advantage of that energy to push air into a building on the windward (upwind) side and cause it to exit on the leeward (downwind) side. Natural ventilation of this type can reduce the fan energy required for moving air inside a building, particularly in a large open-plan office. Restrictions on internal air movement will defeat the purpose of natural ventilation, which requires that engineers and architects make this decision early in the building design process. Because it's so difficult to get just the right wind conditions for natural ventilation, most buildings of this type use a mixed-mode approach, installing fans to help move air in stagnant wind conditions. The figure below shows how natural ventilation might work along with a "stack effect," where heated air rises out the top of a building,

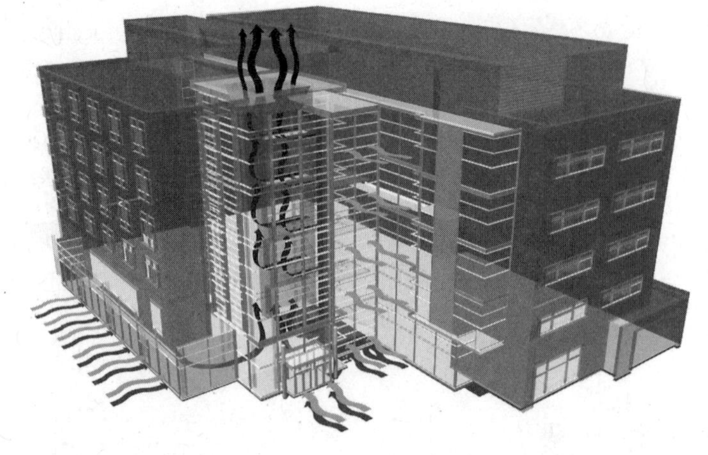

Natural ventilation can complement standard ventilation strategies and reduce fan energy use in larger buildings, as shown for a new engineering building at Portland State University, designed by Zimmer Gunsul Frasca Architects.

with cooler air coming in below. This is a useful approach, since most office buildings require cooling year-round.

In cooler climates ventilation systems need to be coupled with heat-recovery systems, so that heated air leaving the building can give up some of its energy to the incoming air, thus avoiding some of the "energy penalty" of increased ventilation.

Views of the Outdoors

Looking out a window at the outdoors is so fundamental to the human experience that we often fail to notice its importance. Some years ago I saw research data that showed that people who sat farthest from the window in office buildings had higher rates of physiological and psychological illness. Think of a modern office building and a "cube farm" stretching from the windows to the interior core. People in the cubicles farthest from the window can't see outdoors without getting up and moving around (which of course most people do, even if they don't know why). In a green building people should be able to see outdoors from 90% of all regularly occupied spaces (this might exclude some interior conference rooms, for example). In recent years architects have designed offices on the perimeter to have

© Assassi 2006, BNIM Architects

The Lewis and Clark State Office Building in Jefferson City, Missouri, designed by BNIM Architects, is a LEED Platinum-certified building with abundant views of the outdoors.

clear glazing, so anyone can see through them to the outdoors, or to locate corridors along the building perimeter, so everyone can see outside.

When I learned of this requirement in the LEED system, I pondered its importance. But if you think for a moment, human beings have evolved for more than two million years in intimate connection with the outdoors, moving indoors only to eat or sleep. Our entire sensory apparatus, our entire psychology is based on connection with nature; we don't need scientific research to tell us this, it's so obvious. Only in the last 150 years have we begun to spend so much of the daylight hours indoors. Even 100 years ago, 95% of the US population was engaged mostly in agriculture, moving indoors and outdoors frequently.

For most people modern life doesn't allow the luxury of being outdoors much of the day. Have you ever spent an entire day in a windowless conference hall and felt more than a little crazy from lack of connection to daylight and the weather outdoors? We should adopt the more sensible European codes stipulating that no worker can be more than 10 meters (33 feet) from a window, to allow for natural daylighting to penetrate (and for people to see outside). Also, typical cubicle heights need to be reduced to 42 inches or less so that everyone can have a view without getting up from their desks.

Volatile Organic Compounds (VOCs)

Volatile Organic Compounds (VOCs) are an entire class of carbon-based chemicals that give off vapors at normal room temperatures. Thousands of products emit VOCs, including paints and lacquers, paint strippers, adhesives and sealants, carpets and carpet backing, cleaning supplies, pesticides, building materials and furnishings, office equipment (copiers and printers), graphics and craft materials, and permanent markers.

When I was growing up, our family got a new car every three to five years. It was a real treat to sit in the new car and inhale new-car smell, an odor that disappeared after a few months. Of course, as kids we didn't realize that we were breathing several toxic chemicals, including toluene, a suspected carcinogen! For years I found auto-supply stores selling spray cans of new-car smell! VOCs are now regulated by air-quality management districts because they contribute to ground-level ozone.

Nowadays with so many people having environmental allergies, it just makes good sense to reduce the level of VOCs in buildings to which people are exposed, particularly in five major categories:

- Paints and coatings.
- Adhesives and sealants.
- Carpets.
- Composite wood and agrifiber products.
- Furniture and furnishings.

Paints and coatings in green buildings must meet VOC limits established in the Green Seal GS-11 standard, while clear wood finishes, floor coatings and similar substances have their own special standards. Many VOC concentrations are regulated by local air pollution control agencies, since they eventually escape a building through the ventilation system and become an ingredient in ground-level ozone.

High-VOC levels are often found in general construction adhesives (think of anything that comes in a tube), flooring and fire-stopping adhesives, caulking, duct sealants and plumbing adhesives. There are also aerosol adhesives, carpet pad adhesives and ceramic tile adhesives with high-VOC levels.

Carpets and carpet cushions are also sources for VOCs in buildings. In green buildings they must be certified under the Carpet and Rug Institute's Green Label Plus and Green Label programs.

Composite wood and agrifiber products in green buildings must be free of any added urea-formaldehyde resins. This category includes particleboard, medium-density fiberboard, plywood, wheatboard, strawboard and door cores.

Furniture and furnishings are also sources of VOCs. Try to buy a piece of furniture that doesn't have pressed wood fibers soaked with smelly urea-formaldehyde, and you'll understand. Now think about all the people in offices subject to the off-gassing of formaldehyde from new furniture!

Water Conservation

The lament of the Ancient Mariner, "Water, water, everywhere, but not a drop to drink," is echoed in many parts of the world but fortunately not yet in this country. However, looming water shortages in certain parts of the country, such as the Southwest and West, add urgency to the design of water conservation systems for buildings and developments. In addition many are becoming aware of the link between energy use and water use: in the West, considerable water use goes for the cooling needs of thermal (coal) electric power plants. In other words, as this area grows and electricity demand increases, water supplies come under double pressure. Nationwide, freshwater use for power production amounts to about 200 *billion* gallons per day.

Just as the building stock will be 75% new or renovated in 2035, compared with 2005, with all of the implications that has for energy use, by starting now, we can dramatically reduce overall water use for buildings, landscaping and neighborhoods by employing aggressive strategies to reduce consumption of potable water, following the mantra "Reduce, reuse, recycle." Water conservation in buildings involves *reducing* fixture demand, from conventional toilets using 1.6 gallons per flush (gpf) to low-flow toilets using 1.28 gpf (a 20% reduction), then to lower-flows at 1.12 gpf (a 30% reduction), perhaps even to 0.8 gpf (dual-flush toilets on the low setting).

We can eliminate the use of water for flushing urinals entirely, by using water-free urinals. We can *reuse* graywater (wastewater from sinks and showers, for example) for flushing toilets, and we can use onsite wastewater treatment in buildings to provide *recycled* water for such uses as toilet flushing and cooling tower makeup water. We can use both graywater and municipally treated wastewater for landscape irrigation, as well as reducing water demand through xeriscaping strategies.

The key is to manage the entire water cycle, starting with what's free (rainfall) and trying to get as many uses out of that water as possible. For example, if you could reuse 80% of all water used in flushing toilets, you'd get the use of the same water five times. Then, even in a low-rainfall climate, you could have a major impact on water use. Becasue rainfall varies in the US from less than 8 to 12 inches per year in the desert regions to 36 inches in such places as Portland and Chicago, to nearly 50 inches in Orlando, it's important to look at each project's water resources as a starting point for developing design strategies.

The economics of water pricing in urban areas favors looking at the full system costs, including meter charges and connection fees. Reclaiming all the rainwater from a site, so that no storm-sewer connection is needed, can often result in savings that are greater than the cost of rainwater catchment and treatment. The same holds for onsite sewage treatment, particularly if the treated wastewater can be used for toilet flushing, cooling-tower makeup water and irrigation without ever leaving the project site. Why not take a more detailed and expansive look at the opportunities for 40%, or better, water conservation in your next project?

Water-free Urinals

Urinals waste more than 150 billion gallons of fresh water per year, equivalent to the water use of 1,500,000 homes, at an average use of 300 gallons per day per home.[157] The average urinal installed since 1992 uses 1 gallon per flush, which is the code requirement, based on the 1992 federal Energy Policy Act. (Overall, the average might be closer to 2 gallons per flush for all urinals now installed.) Think of more than 78 million men at work,[158] making an average of three flushes per day, five days a week, most of them older urinals using 2 to 3 gallons per flush, just to flush away a liquid that's sterile and more than 99% water.

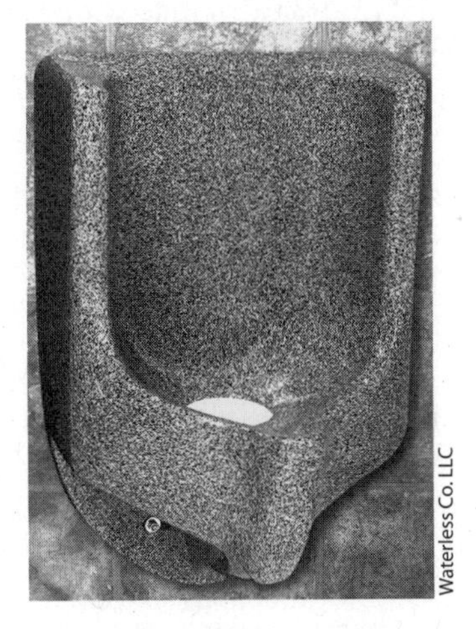

Waterless Co. LLC

Falcon Waterfree Technologies

Two types of water-free urinals.

The design of water-free urinals includes an oil seal below the drain, which prevents sewer gases from rising up (one of the purposes of the flush and the drain) into a bathroom. The seal has to be changed periodically. According to one manufacturer:

> This pleasant-smelling sealant liquid trap provides an airtight barrier between urine and the restroom to prevent odors from escaping the drain, but allows urine to pass through because it is lighter than water. Urine immediately penetrates the sealant liquid and flows to the drain. Uric sediment is collected by the cartridge, leaving an odor-free environment, clean pipes and absolutely no water waste.[159]

Basically, water-free urinals work just fine in situations where there is a large, often anonymous population of users, such as office buildings, restaurants, airports, schools, stadiums and theaters. With proper design and installation, routine maintenance (including quarterly treatment-cartridge replacements) and a little signage to tell users what's going on, water-free urinals work just fine, reducing overall water consumption in buildings by up to 40,000 gallons per year per urinal.[160]

Water-free urinals are used in such places as the Jackie Gleason Theater in Miami Beach; the Evergreen State College in Olympia, Washington; the Harold Washington Social Security Center in Chicago; the twin Petronas Towers in Kuala Lumpur, Malaysia; and the Jimmy Carter Presidential Library, in Atlanta, Georgia.[161] Over the long run, widespread adoption of water-free urinals will also help reduce future infrastructure development costs by reducing water demand and sewage generation. From the standpoint of economics, water-free urinals, either in new buildings or in renovations, pay for themselves in water savings in a relatively short period of time.

Wetlands, Constructed

Constructed wetlands are an integral part of stormwater management and onsite sewage treatment solutions for green building projects. When there is enough land available, such as in suburban office park development, these systems function both to hold stormwater and to use treated sewage

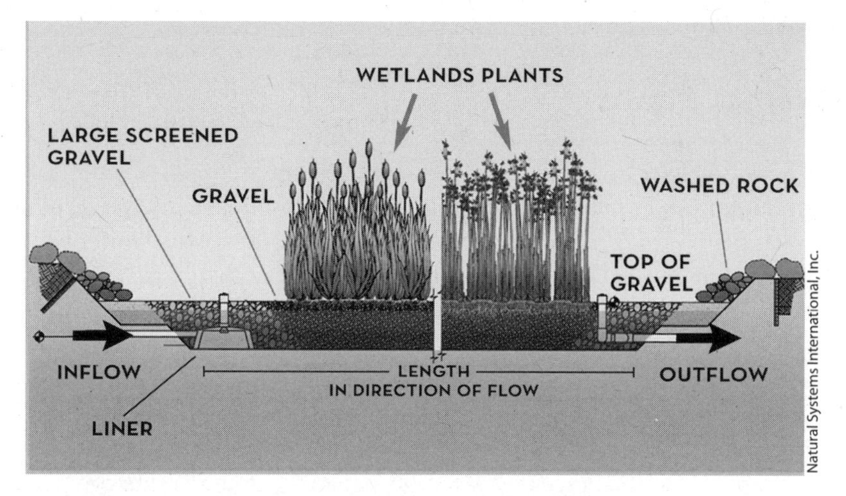

Schematic design of constructed wetlands for sewage treatment.

to supply nutrients to aquatic plants and animals. As most ecology students learn, wetlands are the most productive ecosystems in the world, straddling the intersection of land and water. They are also nature's filters, helping to remove sediments and heavy metals, as well as oil and grease, from stormwater and transforming sewage nutrients into life-giving carbon, nitrogen, phosphorus and trace minerals. Properly designed, constructed wetlands can provide wildlife habitat and open space for green building projects, as well as a place for school tours and outdoor environmental education.

Constructed wetlands can have either surface or subsurface flows. Surface flows resemble marshes, while subsurface flows support a wide variety of plants by supplying them with nutrients. Plantings of reed beds are popular in European constructed wetlands, and plants such as cattails, sedges and bulrushes are used worldwide.[162] As of 2004 more than 5,000 constructed wetlands had been built in Europe, and more than 1,000 were in operation in the US.[163]

Constructed wetlands may be cheaper to build than traditional sewage and stormwater treatment plants, have lower operating and maintenance costs and can handle widely varying volumes of wastewater. For example, the Tres Rios constructed wetlands is a 7-mile, 480-acre riparian corridor in the southwestern part of Phoenix, at the conjunction of the Salt and Gila rivers. As a wildlife sanctuary and place of repose in an increasingly dense urban area, Tres Rios not only treats sewage from the 91st Avenue treatment plant, but provides many public-use benefits for the Phoenix metropolitan area.[164]

Wind Power

The wind blows everywhere on the planet. You might think, why not harness it for power in buildings? Typically, average annual wind speeds of 11 miles per hour are needed for commercial applications, while lower wind speeds can be used for water pumping and battery charging. Now, most cities are not built in places where the average wind speed (every hour of the year) is 11 mph; that's not really very comfortable. So the best wind resources are located away from cities, and the most cost-effective wind power comes from large-scale wind farms that feed the power generated into the electric power grid. On a good site with modern equipment, wind energy costs in the range of 5 to 8 cents per kilowatt hour, quite competitive with most new power sources.[165]

In fact wind power is the largest source of new renewable energy in the US, with nearly 2,500 megawatts (MW) of capacity coming online and feeding into the nation's utility grid in 2006 alone, representing an investment of $4 billion. According to the American Wind Energy Association, in 2006:

> New wind farms boosted cumulative US installed wind energy capacity by 27% to 11,603 MW, the equivalent of 11 large coal-fired electric power plants. Wind energy facilities currently installed in the US produce about 31 billion kilowatt hours annually, enough electricity to serve 2.9 million homes.[166]

Some people advocate placing wind-turbine generators on top of buildings, but this is less cost-effective than in remote locations. First of all, these systems have to be small, which costs more per unit of capacity. Second, wind power increases with height, so taller buildings are better bets than smaller buildings. Third, in cities, upstream buildings can spoil the wind for downstream buildings, so the wind resource has to be evaluated with that in mind, for both existing and whatever future buildings might be erected in your area. Fourth, wind power increases at the *cube* of the wind speed: for example, a site that has 20% more wind speed (for example, 12 mph average vs. 10 mph) will deliver 73% more annual wind power! This basic characteristic of wind power means that only the windiest locations in the windiest cities are good candidates for this power. Fifth, wind power introduces new structural design considerations for buildings.

As with photovoltaics, there are good reasons other than raw economics to consider small wind systems for the tops of buildings, or for building sites, if they produce some energy and operate safely. First, they are very visible and therefore advertise the building's commitment to renewable energy. Second, they can be attractive visual elements, in this case a kinetic sculpture, that architects and building owners might want to use. Third, they provide a teaching tool about renewable energy for schools and environmental education centers. Nevertheless, internationally known wind expert Paul Gipe says, "Mounting wind turbines — of any kind — on a building is a very bad idea. I've yet to see an application where this has worked or will likely work."[167]

Xeriscaping

Outdoor water uses, primarily for landscaping, consume an estimated eight billion gallons per day in the US, perhaps as much as one-third of all water use.[168] "Xeriscaping" is a well-used term for water-conserving landscaping, the prefix denoting dry. Another term might be "natural landscaping." The essential feature of xeriscaping is to employ regionally appropriate plants and planting techniques (such as mulching) that reduce or eliminate water use except from normal precipitation in the area. If you've ever seen sprinklers on in the midst of a rainstorm, or broad expanses of green lawn highlighting a public building or major office complex in the desert, you'll know something is amiss in our understanding of how to minimize the environmental impacts of landscaping practices. Fortunately, xeriscaping is a major movement today among landscape architects.

Where I live in Tucson, I would be shocked to see a green lawn in front of a new home in a development. In the Tucson area, single family residents use 30 to 50 percent of their water outdoors for landscape watering, swimming pools, spas, evaporative cooling and other such uses.[168] The reality of living in a desert has finally caught the popular imagination; as a result, people plant a lot of cactus, mesquite and Palo Verde trees, varieties of succulents and desert wildflowers, living with the seasonality of the dry landscape.

Planting native and adapted species, a typical practice of xeriscaping, has the added advantage of providing food, habitat and shelter for many local birds, insects and small mammals that have evolved in the region. Most state extension services provide information on plant selection for xeriscaping, and most landscape architects know this approach. The harder part is to find someone expert enough in local native plants to introduce and keep them growing in a xeriscaped environment.

Technically, for green buildings xeriscaping can be accompanied by temporary irrigation for the first year, until plants are established. Many practitioners also group plants with similar water requirements, especially in larger areas where different amounts of rainwater can be directed into swales, ponds and waterways. Outside of areas with mild temperatures and abundant rainfall, you will not see much turf in a xeriscape.

The first step in evaluating a xeriscape design is to really know the site: sun, wind, rain, existing vegetation, topography, orientation (south-facing or north-facing, for example) and soils. Plants need to be chosen carefully for their water requirements, shade and sun tolerance, food and shelter value to native wildlife. Group plants in ways that make sense in their natural environment. For example, in the desert young saguaro cacti often grow in the shade of older trees such as mesquite. Put in some mulch and a simple watering system the first year to get the plants established, then let nature take its course.

Zen

My wife and I have an ongoing discussion about how "perfect" things need to be around the home. She likes to have everything "just right." I keep telling her to consider instead the Zen concept of "wabi-sabi," which appeals more to me. Wabi-sabi acknowledges three simple realities: nothing lasts, nothing is finished and nothing is perfect.[169] It implies that how one chooses to look at things is the most important determinant of how satisfied one is with the world. To quote a well-known phrase from yoga philosophy, "The world is as you see it."

Zen is about being satisfied with little things, finding reflections of the cosmos in a bed of sand and gravel, a few well-placed boulders and a sprinkling of natural elements such as grasses. As William Blake wrote in "Auguries of Innocence" more than two centuries ago, you approach the sacred when you manage "to see a world in a grain of sand and a heaven in a wild flower, hold infinity in the palm of your hand and eternity in an hour."

What does this have to do with green buildings? It is this: we need to celebrate what we have achieved, even while we remain insistent on getting better in the future. We need to be aware that each green building is going to have imperfections: things attempted but not achieved, things not attempted that in retrospect could have been accomplished, but didn't fit with the design team's vision or the owner's conception of the project. A Zen approach to green buildings would celebrate also what's special about the place, its particular location on the planet. Perhaps to further this approach, each building could be gifted with a simple haiku, an epigraph at the entrance.

As for a Zen-inspired building, a good example might be the Green Gulch Zen Center Guest House, in the Golden Gate National Recreation Area in Muir Beach, north of San Francisco. According to the project designer, architect Sim van der Ryn:

> The guest house serving the Green Gulch Zen Center is sensitively designed to minimize its visibility and physical impact on site. The octagonal plan and simple elegance of the 12-room guest house reflect the Asian origins of Zen Buddhism. The building's centering geometry is further articulated by a two-story central core, used as a gathering and meeting space.
>
> The design of the guest house supports and respects the

Green Gulch Farm

Green Gulch Zen Center Guest House.

human search for beauty. Handcrafted Japanese joinery is used throughout the house with strict attention to detail. The building incorporates recycled timbers and a passive solar heating system designed to meet human needs without destroying the fabric of the living world. [This] is a truly peaceful place.[170]

Zero-net-energy Buildings

The concept of zero-net-energy buildings and zero-net-energy neighborhoods is quickly capturing the attention of many green building designers and even some developers. A zero-net-energy building would provide all of its own energy on an annual basis from onsite renewable resources or offsite renewable energy purchases. In this way it would still be connected to the grid, providing power when it had a surplus and drawing from the grid when it needed power, such as at night.

This approach typically involves using solar energy for electricity, water heating and space heating and employing such design measures as passive solar design, natural ventilation and operable windows for space cooling (with some electric power assist). In practice complete energy independence is quite achievable at the level of homes and small buildings. Much depends on the local microclimate; yet if one thinks about it, all homes were "zero-net-energy" before the Industrial Revolution, so there are many sources of indigenous architecture for inspiration!

A single-family home in Paterson, New Jersey, certified at LEED for Homes Platinum level in 2006, shows how to move toward the goal of zero-net-energy. Called The BASF Near-Zero Energy Home, it includes expanded polystyrene insulation, polyurethane foam sealants and cool metal roof coatings to reduce energy use 80% below a typical home.[171]

On a larger scale, in March 2006 the World Business Council for Sustainable Development announced that it is forming an alliance to develop zero-energy buildings. They have an ambitious target: by 2050 new buildings will consume zero-net-energy from external power sources and produce zero-net carbon dioxide emissions while being economically viable to construct and operate.[172]

In December 2006 the UK government announced a program for zero-carbon new homes; by 2016 all new homes are to be zero-carbon, with

Gridley & Graves

The BASF Near-Zero Energy Home in Paterson, New Jersey.

a 25% improvement on energy use over current building regulations by 2010 and a 44% improvement by 2013. A 100-home project near London, the BedZED (Beddington Zero Energy Development) set a goal of becoming carbon neutral; they have cut carbon emissions by 56% through energy-efficiency measures and an onsite solar photovoltaics system.[173]

PART III

Resources

Resources

1. Books

Most books are outdated shortly after they are published in this fast-changing field. Nevertheless, there are a few that have some degree of shelf life. You might find them interesting, perhaps life-changing.

Ray Anderson, *Mid-Course Correction*, Peregrinzilla Press, 1998.
This is a classic book chronicling the successful beginning of a corporate paradigm shift through a personal transformation by the CEO. Ray Anderson speaks from the heart, with experience, passion and eloquence.

Janine Benyus, *Biomimicry: Innovation Inspired by Nature*, Harper Perennial, 2002.
This is the bible for all those who think that four billion years of natural evolution might have some lessons for contemporary technologists looking for high-performance systems with less resource inputs.

Penny Bonda and Kate Sosnowchik, *Sustainable Commercial Interiors*, Wiley, 2006.
This is a very recent book for practicing interior designers. Penny Bonda has been widely recognized as a significant force in greening the interior design profession.

Stewart Brand, *How Buildings Learn*, Penguin, 1995.
Widely lauded as a work of genius, this book shows how buildings change and learn over time; it argues for flexibility, a key component of all green building design.

Herbert Dreiseitl and Dieter Grau, *New Waterscapes: Planning, Designing and Building with Water*, Birkhaeuser, 2005.
Dreiseitl is an acknowledged contemporary master of introducing water into landscape architecture, with the broader perspective of an art therapist (his first profession) on how water brings us together and heals the soul. This book updates his earlier work (*Waterscapes*) with his latest projects.

Kari Foster, Annette Stelmack and Debbie Hindman, *Sustainable Residential Interiors*, Wiley, 2006.
Basing their approach on an integrated design process, Associates III, a leading interior design firm in Boulder, Colorado, with expertise in addressing environmental concerns in homes, presents solutions for the residential interior designer.

David Gottfried, *Greed to Green*, WorldBuild Publishing, 2004.
If you want the insider's perspective on the formation and early years of the US Green Building Council, David Gottfried's amazing story of personal and organizational transformation pulls no punches.

Paul Hawken, Amory Lovins and L. Hunter Lovins, *Natural Capitalism: Creating the Next Industrial Revolution*, Little Brown, 1999.
This book is a classic treatment of a wide variety of subjects, all based around the idea of how much we can learn from natural systems and how little we are applying what we already know. This book will reward anyone who wants to understand how to take the next leap in green building design.

David Macaulay and Jason McLennan, *The Ecological Engineer, Volume One: Keen Engineering*, Ecotone, 2005.
This is the first volume in a series profiling the work of the most innovative engineering firms in North America and showing how they produce functionally outstanding structures, systems and technologies. Kevin Hydes, former president of Keen Engineering, wrote the foreword and contributed to the book.

William McDonough and Michael Braungart, *Cradle to Cradle: Changing the Way We Make Things*, North Point Press, 2002.
Not printed on ordinary book paper, this book walks the talk. The authors take us step by step through their reasons for advocating a new industrial paradigm and present great case studies of how they've begun the process for a number of companies.

Jason McLennan, *Philosophy of Sustainable Design*, Ecotone, 2004.
This is an early attempt at a "philosophy" of sustainable design; an eminently readable overview of the key concepts used by practitioners in the field. Respect nature, while making things green and beautiful are key take-aways from this book.

Sandra Mendler, William Odell and Mary Ann Lazarus, *The HOK Guidebook to Sustainble Design*, 2nd Ed., Wiley, 2006.
If there's one essential reference for anyone practicing or wanting to understand the realities of sustainable design, this is it. There are 400 pages of how-to with lots of practical examples.

Ross Spiegel and Dru Meadows, *Green Building Materials: A Guide to Product Selection and Specification*, 2nd Ed., Wiley, 2006.
This is definitely a book for specialists in product selection and specification, but the first part of the book is a great overview of the product selection process, "where the rubber meets the road" in green building.

Alex Steffen, ed., *World Changing: A User's Guide for the 21st Century*, Abrams, 2006.
It's hard to know what to say about this nearly 600-page compendium of everything we know about green solutions, except that you need a copy in your library for reference.

Sim van der Ryn, *Design for Life: The Architecture of Sim van der Ryn*, Gibbs Smith, 2005.
The subtitle gives away the book's message. This is the story of Sim's life work. It is told brilliantly, movingly and from a very personal perspective. Sim's work has influenced thousands of sustainable designers.

Sim van der Ryn and Stuart Cowan, *Ecological Design: Tenth Anniversary Edition*, Island Press, 2007.
A seminal work that illustrates and explicates five core principles of sustainable design, it is a timeless reminder of how much designers have to learn from natural systems and how they should design with the long-term health of people and planet foremost in mind.

Alex Wilson, *Your Green Home*, New Society Publishers, 2006.
This is a great primer for anyone considering building a new home and wanting to make it as green as possible. Alex probably has the best overview of these topics of anyone around.

Alex Wilson and Mark Piepkorn, eds., *Green Building Products: The Green Spec Guide to Residential Building Materials*, 2nd Ed., New Society Publishers, 2006.
This is the single-best desktop reference on actual products for anyone designing new green homes or commercial buildings, thoroughly vetted by the publishers of *Environmental Building News*.

James Wines, *Green Architecture*, Taschen, 2000.
James Wines is a professor of architecture and someone who can tell stories with the best of them. This book is a beautifully illustrated overview of the state of the art of green design through the early 2000s.

Jerry Yudelson, *Developing Green: Strategies for Success.*, National Association of Industrial and Office Properties, 2006, www.naiop.org, with CD of case studies attached.

This is the best introduction to the business case for green buildings, written for developers. Includes case studies of green developments submitted for the first NAIOP Green Development of the Year award in 2005.

2. Publications

It's hard to keep up with the proliferation of green building magazines and related publications. Here are a few publications I read on a regular basis and find valuable for staying in touch. Most are available both in hard copy and electronic versions, so if you're averse to having too much paper around, you can keep up with the news via electronic editions

Building Design & Construction, www.bdcmag.com
BD&C's editor Rob Cassidy is one of the authoritative voices in the industry. Written primarily for "Building Team" practitioners, *BD&C* is eminently accessible to anyone.

Buildings Magazine, www.buildings.com
Buildings magazine provides a good introduction to the practical side of building design, construction and operations. Good coverage of specialty topics in the industry.

Consulting-Specifying Engineer, www.csemag.com
If you're an engineer working with green buildings, this publication provides regular coverage of the engineering issues in green building design and construction.

eco-structure, www.eco-structure.com
Eco-structure is the most illustrated of the trade magazines covering the green building industry. Good case studies and a broad selection of topics make it a good read for keeping up.

Environmental Building News, www.buildinggreen.com
EBN is simply the best-edited and most relevant publication for green builders. The monthly feature stories keep you abreast of emerging issues in green building design, construction and operations.

Environmental Design & Construction, www.edcmag.com
Now ten years old, ED&C provides first-class editorial coverage of relevant issues, along with well-written cases studies of leading green building projects.

Green@Work Magazine, www.greenatworkmag.com
A bi-monthly magazine that provides a good overview of sustainability in the workplace, including extensive coverage of green buildings.

Green Builder Magazine, www.greenbuildermag.com
This is the monthly green building magazine for the 95,000 members of the National Association of Home Builders. Essential reading if you want to know what this audience is learning about green buildings.

Green Source Magazine, www.construction.com/greensource
From the publishers of *Engineering News-Record* and *Architectural Record*, authoritative publications in their field, this quarterly is edited by the team at Environmental Building News. The case studies are the best written you will find anywhere.

Home Energy Magazine, www.homeenergy.org
A fairly technical magazine on energy-efficiency home design and home improvements, this journal is accessible to the serious homeowner as well as anyone with good technical skills.

Journal of Greenbuilding, www.collegepublishing.us/journal.htm
This is the first serious academic journal in the field, published quarterly and evenly divided between research articles and review articles of key issues in green building.

Metropolis, www.metropolismag.com
If you want to know what's going on in the broader world of design, this monthly is a must-read.

Renewable Energy World, www.renewable-energy-world.com
This is a lavishly illustrated, technically accurate (without being off-putting) journal from the UK that covers the spectrum of renewable energy developments. Best of all, it's free.

Solar Today, www.solartoday.org
This official publication of the American Solar Energy Society is written for a general audience; you can even find it at the checkout counter of natural foods stores.

Sustainable Industries Journal, www.sijournal.com
This monthly provides great coverage of the West Coast's developments across a wide range of sustainable industries, including green building. Short articles, easy to read.

3. Websites

Green Building Initiative, www.thegbi.org
This is the official website for the Green Globes rating system. At this site, you

may register and download a trial version of the system for use in one of your projects.

GreenBuzz, www.greenbuzz.com
A good website for a broader view of the sustainable business movement.

IGreenBuild, www.igreenbuild.com
This is a good overview website of the business and product side of the green building movement.

National Association of Homebuilders, Model Green Home Guidelines, www.nahb.org
At this site, you may download the green home rating system preferred and currently used by most homebuilders and homebuilder associations.

Sustainable Buildings Industry Council, www.sbicouncil.org
The SBIC is a leading national educational organization focused heavily on homes and high-performance schools. Its residential green building guidelines came out in a fifth edition in 2007.

US Green Building Council, www.usgbc.org
The USGBC website is the premier website not only for the organization but for news and happenings in the broader field of green buildings. If a trend has legs, you'll find it here. You can download copies of all LEED rating systems and also search for LEED-registered and certified projects.

US Dept. of Energy, High-Performance Bldgs. Database, www.eere.energy.gov/buildings/database/
This is a great site for detailed case studies of nearly 100 green building projects.

US Environmental Protection Agency, ENERGY STAR program, www.energystar.gov
The essential site for learning about ENERGY STAR ratings for products, homes and buildings, with lots of great free tools you can use to assess the energy performance your building, compared with a database of more than 3,000 other buildings' actual energy use.

World Changing, www.worldchanging.com
Emerging innovations and solutions for building a brighter green future; an essential site if you want to know what's going to be a mainstream concern in short order.

Endnotes

Preface

1. US Green Building Council data, January 2007.

Chapter 1

1. *City Mayors: Largest cities in the world* [Cited April 3, 2007]. citymayors.com/statistics/largest-cities-2007.html
2. *Cradle to Cradle C2C Home Competition Winner* [Cited April 3, 2007]. cradletocradlehome.com/wst_page2.html
3. [Cited April 3, 2007]. precaution.org/lib/pp_def.htm
4. *The Hannover Principles: Design for Sustainability* [Cited April 3, 2007]. mcdonough.com/principles.pdf
5. *Environmental and Energy Study Institute. Renewable Energy and Energy Efficiency Fact Sheets* [Cited April 3, 2007]. eesi.org/programs/energy andclimate/EEREFactSheetsIndex.html
6. [Cited April 3, 2007]. architecture2030.org

Chapter 2

1. Data from USGBC, January 24, 2007.
2. [Cited February 25, 2007]. cagbc.org
3. Brownlie, Mark. *Trend Watching* [Cited April 3, 2007]. According to the *KPMG International Survey of Corporate Sustainability Reporting 2002*, approximately 45% of the Global Fortune 250 now produces some type of social, environmental, corporate citizenship or sustainability report. October 2003. greenbiz.com/news/columns_third.cfm?NewsID=25699
4. construction.com and *Architectural Record*, December 27, 2005.
5. *Introduction to LEED and Green Building* [Cited April 3, 2007]. usgbc.org/showfile.aspx?DocumentID=742
6. gghc.org
7. cagbc.org
8. Interview in *Urban Land* magazine, January 2007, p. 118.

Chapter 4

1. GreenBiz.com. *Swiss Re Offers Employee Rebate to Reduce Carbon Footprint* [Cited January 9, 2007]. greenbiz.com/news/news_third.cfm?NewsID=34400

2. [Cited April 3, 2007]. resource-solutions.org
3. [Cited April 3, 2007]. greenercars.org
4. *Cities Working Together to Protect Our Air Quality, Health and Environment: A Call to Action* [Cited April 3, 2007]. seattle.gov/mayor/climate/PDF/USCM_6-page_Climate_Mailing_ALL.pdf
5. *Clinton Global Initiative* [Cited April 3, 2007]. attend.clintonglobalinitiative.org/home.nsf/pt_cmt_topic?open&cat=climate
6. [Cited April 3, 2007]. greenprintdenver.org
7. [Cited April 3, 2007]. dsireusa.org
8. *Making Solar Power Mainstream* [Cited April 3, 2007]. environmentcalifornia.org/energy/million-solar-roofs/fact-sheet
9. aashe.org
10. [Cited April 3, 2007]. asusustainability.asu.edu
11. Leith Sharp, Harvard University. Personal communication, January 17, 2007.

Chapter 5

1. [Cited April 3, 2007]. architecture2030.org
2. [Cited April 3, 2007]. architecture2030.org
3. Sierra Club. *Bicycle Commuting* [Cited April 3, 2007]. sierraclub.org/e-files/bike_commuting.asp. Also, *Outside Magazine* outside.away.com/outside/destinations/200608/best-outside-towns-2006-6.html [Cited 3/30/0].
4. *Total Midyear Population for the World* [Cited December 30, 2006]. census.gov/ipc/www/worldpop.html
5. *Fundamentals of Physical Geography* [Cited April 3, 2007]. physicalgeography.net/fundamentals/5d.html
6. Flannery, Tim. *The Weather Makers: The History and Future Impact of Climate Change.* Text Publishing, 2005.
7. "Chicken Fat Could Soon Find Use in Biofuel." *Arizona Daily Star.* January 8, 2007, p. D1.
8. *Contact Us: FAQs* [Cited January 8, 2007]. vw.com/contactus/faqs.html#5.1
9. Cook, John. *Biodiesel On Sale at West Seattle Safeway* [Cited February 23, 2007]. seattlepi.nwsource.com/business/304443_biodiesel21.html?source=rss
10. See note 7.
11. Wilson, Edward O. *Biophilia.* Harvard University Press, 1984.
12. h-m-g.com
13. *Windows and Offices: A Study of Office Worker Performance and the Indoor Environment* [Cited April 3, 2007]. h-m-g.com/downloads/Daylighting/A-9_Windows_Offices_2.6.10.pdf
14. [Cited April 3, 2007]. h-m-g.com/downloads/Daylighting/A-5_Daylgt_Retail_2.3.7.pdf
15. Reference material from AIG Global Real Estate for Atlantic Station.

16. Development Center for Appropriate Technology [Cited April 3, 2007]. dcat.org
17. *Sustainability and the Building Codes* [Cited January 8, 2007]. networkearth.org/naturalbuilding/codes.html
18. Interface Engineering. *Engineering a Sustainable World*, 2005.
19. [Cited April 3, 2007]. glumac.com
20. "China Cashes in on Global Warming." *Wall Street Journal*, January 8, 2007, p. A11.
21. [Cited April 3, 2007]. makower.typepad.com/joel_makower/2007/01/is_carbon_neutr.html
22. [Cited April 3, 2007]. shawcontractgroup.com
23. David Hudnall, Louisiana Pacific Corp. Personal communication, December 2006.
24. US Green Building Council. *LEED for New Construction and Major Renovation, Version 2.2 Reference Guide*. September 2006, p. 368.
25. *Cost-Effectiveness of Commercial-Buildings Commissioning* [Cited April 3, 2007]. eetd.lbl.gov/Emills/PUBS/Cx-Costs-Benefits.html
26. Deane, Michael. *Can Turner Construction Make 50% Waste Diversion Standard Practice in a Major Construction Company*, furnished by the author, 2005.
27. [Cited April 3, 2007]. fypower.org/res/tools/products_results.html?id=100123
28. [Cited April 3, 2007]. coolroofs.org
29. *Cool Roofs* [Cited April 3, 2007]. fypower.org/res/tools/products_results.html?id=100123
30. Dennis Wilde, GerdingEdlen Development. Personal communication, spring 2006.
31. Hawken, Paul, Amory Lovins and L. Hunter Lovins. *Natural Capitalism*. Little, Brown & Company, 1999, p. 115.
32. Matthiessen, Lisa Fay and Peter Morris. *Costing Green: A Comprehensive Cost Database and Budgeting Methodology* [Cited April 3, 2007]. davislangdon.us/pdf/USA/2004CostingGreen.pdf. July 2004.
33. McDonough Braungart Design Chemistry, LLC. *Cradle to Cradle Certification Criteria* [Cited December 31, 2006]. mbdc.com/docs/CertificationCriteria.pdf
34. *Interface Launches Mission Zero* [Cited December 31, 2006]. interfaceinc.com/pdfs/InterfaceLaunchesMissionZero.pdf
35. Anderson, Ray. *Mid-Course Correction, Toward a Sustainable Enterprise: The Interface Model*. Peregrinzilla Press, 1998.
36. *European Union Outpaces United States on Chemical Safety* [Cited January 3, 2007]. huliq.com/4526/european-union-outpaces-united-states-on-chemical-safety
37. Selen, Henrik and Stacy VanDeveer. "Raising Global Standards," *Environment*, 4810, December 2006, pp. 6–17. Reprinted with the permission of the Helen Dwight Reid Educational Foundation. Published by

Heldref Publications, 1319 18th Street NW, Washington, DC 20036-1802. www.heldref.org Copyright© 2006.
38. *Building Investment Decision Support (BIDS)* [Cited January 5, 2007]. aia.org/SiteObjects/files/BIDS_color.pdf P. 22.
39. *Windows and Offices: A Study of Office Worker Performance and the Indoor Environment-CEC PIER 2003* [Cited April 3, 2007]. h-m-g.com /projects/daylighting/summaries%20on%20daylighting.htm
40. Al Nichols, Al Nichols Engineering, Tucson. Personal communication.
41. Peters, Tom. *Re-Imagine!*. DK Adult, 2003; and *Design*. DK Adult. 2005.
42. Brand, Stewart. *How Buildings Learn*. Penguin, 1995.
43. Wackernagel, M., and W. Rees. *Our Ecological Footprint: Reducing Human Impact on the Earth*. New Society Publishers, 1996.
44. *National Footprints* [Cited April 3, 2007]. footprintnetwork.org/gfn_sub.php?content=national_footprints
45. *Ecological Footprint: Overview* [Cited April 3, 2007]. footprintnetwork.org/gfn_sub.php?content=footprint_overview
46. naturalstep.org
47. SERA Architects. *A Natural Step Case Study* [Cited April 3, 2007]. Winter 2005. serapdx.com
48. Kutscher, Charles F. *Tracking Climate Change in the U.S.* [Cited April 3, 2007]. ases.org/climatechange/climate_change.pdf
49. Ross Spiegel, Fletcher Thompson Architects/Engineers. Personal communication, January 23, 2007.
50. Bender, Tom. *Building with the Breath of Life*. Fire River Press, 2000, p. 68.
51. Environmental Health Center. *Formaldehyde* [Cited April 3, 2007]. nsc.org/EHC/indoor/formald.htm. See also, U.S. Department of Labor osha.gov/SLTC/formaldehyde/index.html [April 13, 2007]. hermanmiller.com/CDA/SSA/Product/1,1592,a10-c440-p225,00.html [Cited April 3, 2007].
52. Green Building Initiative. *News* [Cited January 24, 2007]. thegbi.org/gbi/news_120806_sbic.asp
53. *Green Globes Design v. 1: Post-construction Assessment* [Cited April 3, 2007]. thegbi.org/greenglobes/pdf/GreenGlobesDSTU.pdf. P. 5.
54. thegbi.org/residential. [Cited January 24, 2007].
55. *Green Building Rating Systems* [Cited January 24, 2007]. thegbi.org/gbi/Green_Building_Rating_UofM.pdf
56. [Cited April 3, 2007]. thegbi.org/greenglobes/pdf/GreenGlobesDSTU.pdf
57. *Green Guide for Health Care Newsletter*. May 2006. gghc.org
58. *The Business Case for Green Buildings* [Cited January 24, 2007]. healthcaredesignmagazine.com/CleanDesign.htm?ID=5148
59. Ibid.
60. *NAHB's Model Green Home Building Guidelines* [Cited April 3, 2007]. nahb.org/publication_details.aspx?publicationID=1994§ion ID=155

61. *SGS Solar System Description* [Cited January 7, 2007]. greenwatts.com/pages/SolarStats/SolarDescr.html
62. green-e.org
63. *Verification Report Reveals Significant Increase in Certified Renewable Energy Sales in 2005* [Cited April 3, 2007]. resource-solutions.org /where/pressreleases/2006/Verification_Report_Release.12.04.06.htm
64. [Cited April 3, 2007]. b-e-f.org/GreenTags
65. *Ten Criteria for Evaluating Green Building Materials* [Cited January 6, 2007]. heartlandgreensheets.org/10criteria.html#7
66. See the Green Spec directory, described in the Resources section, for a listing of more than 2,000 green products, their characteristics and where to get them.
67. *A Conversation with William McDonough* [Cited January 10, 2007]. Spring 2006. darden.virginia.edu/batten/batten_briefings/Articles/battenbrf_spring_06_mcdonough.pdf
68. *High-Performance Buildings Database* [Cited April 3, 2007]. US Department of Energy. eere.energy.gov/buildings/highperformance/
69. [Cited April 3, 2007]. eia.doe.gov/emeu/cbecs/contents.html
70. [Cited April 3, 2007]. powerlight.com/products/suntile_features.php
71. *California Builder Magazine* [Cited January 24, 2007]. cabuilder.com/internal.asp?pid=272
72. *Grupe Earns Certification for Meeting Quality Green Building Standards at Carsten Crossings in Rocklin* [Cited April 3, 2007]. grupe.com/communities/carsten/news_grupe_LEED_cert.pdf
73. Benton, Joe. *Hybrid Sales Drop with Gas Prices* [Cited January 6, 2007]. consumeraffairs.com/news04/2007/01/hybrids_trends.html
74. *About: Hybrid Cars* [Cited January 6, 2007]. hybridcars.about.com/od/hybridcarfaq/f/hybridtaxamount.htm
75. *The Greenest Vehicles of 2007* [Cited January 6, 2007]. greenercars.com/12green.html
76. *Oil Prices: A Pause, Then Up, Part II* [Cited April 3, 2007]. online.barrons.com/article/SB116077947892592476.html?mod=article-outset-box
77. *Interface Showroom Achieves LEED-CI Platinum* [Cited April 3, 2007]. bentleyprincestreet.com/cultures/en-us/bpc/news/pg_leed-ci_platinum.htm
78. Cook+Fox Architects. Personal communication of project description. cookplusfox.com
79. *Society for Neuroscience Wins Gold Certification Award from U.S. Green Building Council for New Office Space* [Cited April 3, 2007]. sfn.org/index.cfm?pagename=news_112806
80. *Press Releases* [Cited April 3, 2007]. hines.com/press/releases/9-27-06.aspx
81. *News Releases* [Cited December 31, 2006]. shareholder.com/lry/investors/ReleaseDetail.cfm?ReleaseID=214276&ReleaseType=VABeach

82. venture-magazine.com/content_archives/Fall06/index.html [Cited December 31, 2006].
83. Sims, Stephanie. *LEEDing the Way* [Cited December 31, 2006]. gerdingedlen.com/project.php?id=62
84. *Alley24 East* [Cited December 31, 2006]. vulcanrealestate.com/TemplateSuccessStories.aspx?contentId=118
85. [Cited December 31, 2006]. uli.org/AM/Template.cfm?Section=GreenTech1&Template=/MembersOnly.cfm&ContentID=37654
86. *The New Orleans Principles* [Cited April 3, 2007]. p. 4. green_reconstruction.buildinggreen.com/documents.attachment/305068/NewOrleans_Principles_LowRes.pdf
87. Mouzon, Steve. *Cottages to Call Home* [Cited January 11, 2007]. igreenbuild.com/cd_2115.aspx
88. *Affordable Green Housing* [Cited April 3, 2007]. nrdc.org/cities/building/fhousing.asp
89. hermanmiller.com/CDA/SSA/Product/1,1592,a10-c440-p225,00.html [Cited April 13, 2007].
90. US Green Building Council. *LEED-EB Rating System* , p. 3. [Cited January 10, 2007]. usgbc.org
91. *LEED for Neighborhood Developments Rating System: Pilot Project Rating System* , pp. 4–5 [Cited February 7, 2007]. usgbc.org
92. *BEES 3.0* [Cited January 11, 2007]. bfrl.nist.gov/oae/software/bees.html
93. Scientific Application International Corporation. *Life Cycle Assessment: Principles and Practice* [Cited January 11, 2007]. epa.gov/ORD/NRMRL/lcaccess/pdfs/600r06060.pdf
94. Andy Frichtl, Interface Engineering, Inc., Portland, Oregon. Personal communication.
95. *ECANL Conference* [Cited February 25, 2007]. urbanwildlands.org/conference.html
96. *How Energy Is Used in Commercial Buildings* [Cited April 3, 2007]. 2004. eia.doe.gov
97. *The Living Building Challenge* [Cited April 3, 2007]. cascadiagbc.org/resources/living-buildings/living-building-challenge
98. Head, Peter. Presentation at the 2007 Green Cities conference, Sydney, Australia. This development has been featured in a number of news articles in 2006 and 2007.
99. *NW Regional Green Buildings Products Matrix* [Cited April 3, 2007]. sconnect.org/greenbuilding/nwwashingtongreenbuildingmaterialsmatrix
100. "New Toilets Are Going Green to Halt Gallon Guzzling." *Washington Post.* January 21, 2007.
101. *Environmental Building News* [Cited April 3, 2007]. February 2006. buildinggreen.com/articles/IssueTOC.cfm?Volume=16&Issue=2
102. *Technology Basics* [Cited April 3, 2007]. eere.energy.gov/de/microturbines/tech_basics.html

103. Dreiseitl, Herbert and Dieter Grau, eds. *New Waterscapes: Planning, Designing and Building with Water*. Birkhaueser, 2005, pp. 42–45.
104. *Thomas Crapper* [Cited April 3, 2007]. theplumber.com/crapper.html
105. Interface Engineering. *Engineering a Sustainable World*, p. 27.
106. *LEED-EB Project Case Study: JohnsonDiversey Headquarters* [Cited April 3, 2007]. usgbc.org/Docs/LEEDdocs/JohnsonDiversey%20 Narrative%20Case%20Study%20V5.pdf
107. *Project Profile* [Cited April 3, 2007]. usgbc.org/ShowFile.aspx?DocumentID=2058
108. *Press Releases* [Cited April 3, 2007]. usgbc.org/News/PressReleaseDetails.aspx?ID=2783
109. *Ozone Hole* [Cited April 3, 2007]. antarctica.ac.uk/Key_Topics/The_Ozone_Hole
110. Trane Corporation. Personal communication, February 2007.
111. *Paints (GS-11)* [Cited April 3, 2007]. greenseal.org/certification/ standards/paints.cfm. For Green Planet Paints, see greenplanetpaints .com [Cited April 13, 2007].
112. See the classic work on this subject: Kuhn, Thomas. *The Structure of Scientific Revolutions*. University of Chicago Press, 1996.
113. Sobel, Dava. *The Planets*. Penguin, 2006, p. 38.
114. Meadows, Donella. *Leverage Points: Places to Intervene in a System* sustainer.org/pubs/Leverage_Points.pdf
115. Mazria, Edward. *The Passive Solar Energy Book: A Complete Guide to Passive Solar Home, Greenhouse and Building Design*. Rodale Press, 1979.
116. Brown, G.Z., and Mark DeKay. *Sun, Wind & Light: Architectural Design Strategies*, 2nd ed. Wiley, 2000.
117. *Non-point Source Pollution: The Nation's Largest Water Quality Problem* epa.gov/nps/facts/point1.htm. See also, epa.gov/epaoswer/hazwaste /usedoil/index.htm [Cited April 3, 2007].
118. *CE News*. January 2007, p. 11. cenews.com
119. *Permeable Paving* [Cited April 3, 2007]. toolbase.org/Technology-Inventory/Sitework/permeable-pavement
120. Other state incentives for renewable energy and energy efficiency can be viewed at: dsireusa.org
121. Yudelson, Jerry. *Making Photovoltaics Pay Their Way* [Cited January 14, 2007]. bdcnetwork.com/article/CA6390972.html?text=photovoltaics
122. *Genzyme Center* [Cited April 3, 2007]. leedcasestudies.usgbc.org/overview.cfm?ProjectID=274
123. Turner, Cathy. *LEED Building Performance in the Cascadia Region: A Post Occupancy Evaluation Report* [Cited April 3, 2007]. usgbc.org/chapters/cascadia/docs/pdf/POE_REPORT_2006.pdf
124. *Occupant Satisfaction with IEQ in Green and LEED-Certified Buildings* cbe.berkeley.edu/research/briefs-LEED.htm
125. Redrawn from Carnegie Mellon University, Energy Building Informa-

tion Decision Support (eBIDS) studies.
cbpd.arc.cmu.edu/ebids/pages/home.aspx [Cited April 13, 2007].
126. Harrington, Dan. "Rapidly Renewable Revolution." *Eco-Structure.* January–February, 2006, pp. 62–63.
127. *Environment* [Cited January 6, 2007]. homasote.com/enviro.html
128. *Solar Power* [Cited April 3, 2007]. library.thinkquest.org/C004471/tep/en/traditional_energy/solar_power.html
129. US Department of Energy. *Annual Energy Outlook* [Cited April 3, 2007]. eia.doe.gov/oiaf/aeo/pdf/appa.pdf
130. [Cited April 3, 2007]. 25x25.org
131. [Cited April 3, 2007]. Ibid.
132. Kutscher, Charles F. *Tracking Climate Change in the U.S.*, p. 4 [Cited April 3, 2007]. ases.org/climatechange/climate_change.pdf
133. Yudelson, Jerry. *Developing Green: Strategies for Success.* National Association of Industrial and Office Properties (NAIOP), Herndon, VA, 2006. Available at: naiop.org/bookstore
134. Ibid., p. 80.
135. Interface Engineering. *Engineering a Sustainable World.* 2005, p. 19. Available by special order from ieice.com
136. [Cited April 3, 2007].
Z-m-g.com/downloads/Daylighting/day_registration_form.htm
137. Kats, Greg. *Greening America's Schools: Costs and Benefits* [Cited April 3, 2007]. 2006. cap-e.com/ewebeditpro/items/O59F9819.pdf
138. Butti, Ken and John Perlin. *A Golden Thread: 2500 Years of Solar Architecture and Technology.* Cheshire Books, 2006.
139. McLennan, Jason F. *The Philosophy of Sustainable Design.* Ecotone, 2004, p. 4.
140. Steffen, Alex. *World Changing: A User's Guide for the 21st* Century. Abrams, 2006, p. 86.
141. *Wall St. Journal.* January 17, 2007, p. B3.
142. *Sustainable Industries Journal.* March 2007, pp. 24–25. sijournal.com
143. *Ecomagination* [Cited April 3, 2007].
ge.com/en/citizenship/customers/markets/ecomagination.htm
144. *Greenbuild 2007 Chicago November 7–9* [Cited January 26, 2007]. greenbuildexpo.org/DisplayPage.aspx?CMSPageID=4
145. "Double Take." *Archi-tech* magazine. January 2007, p. 34. architechmag.com
146. Lockwood, Charles. "As Green as the Grass Outside." *Barron's.* December 25, 2006, p. 37.
147. Yudelson, Jerry. *Developing Green: Strategies for Success.* NAIOP, 2006, pp. 1–12. naiop.org
148. McDonough, William and Michael Braungart. *Design for the Triple Top Line,* August 2002. mcdonough.com/writings/design_for_triple .htm
149. Gottfried, David. *From Greed to Green.* WorldBuild Books, 2004.

150. *Chapters* [Cited April 3, 2007]. usgbc.org/DisplayPage.aspx?Category ID=24. See also the Canada GBC Web site, cagbc.org/membership _information/statistics.php [Cited April 13, 2007].
151. *Value of Private Nonresidential Construction Put in Place: Seasonally Adjusted Annual Rate* [Cited April 3, 2007]. census.gov/const/C30/nonresidentialsa.pdf
152. Fedrizzi, Rick. *Opening Plenary Remarks by S. Richard Fedrizzi* usgbc.org/Docs/News/openplenaryrick.pdf
153. *Project Overview* [Cited April 3, 2007]. docksidegreen.ca/dockside_green/overview/index.php
154. *LEED for New Construction, Version 2.2 Reference Guide.* USGBC, 2006, p. 92.
155. *The Tower Companies and Lerner Enterprises Moving Corporate Headquarters to New LEED Office Building in Tower Oaks* prnewswire.com/cgi-bin/stories.pl?ACCT=104&STORY=/www/story /01-09-2007/0004502535&EDATE=
156. toweroaks.com
157. *Waterless No-Flush Urinals* [Cited January 21, 2007]. waterless.com/conservation.php
158. *U.S. Department of Labor: Bureau of Labor Statistics* [Cited January 21, 2007]. bls.gov/webapps/legacy/cpsatab1.htm
159. [Cited April 3, 2007]. falconwaterfree.com/how/stage2.htm
160. waterless.com [Cited January 21, 2007].
161. Manufacturers' Websites: waterless.com and falconwaterfree.com [Cited April 3, 2007].
162. [Cited April 3, 2007]. epa.gov/owow/wetlands/watersheds/cwetlands.html
163. *Constructed Wetland* [Cited January 21, 2007]. epa.gov/owow/wetlands/pdf/ConstructedW_pr.pdf
164. *Future: Full Scale Wetlands* [Cited January 21, 2007]. phoenix.gov/TRESRIOS/future.html
165. *Resources* [Cited January 21, 2007]. awea.org/resources/financing/cost.html
166. *Wind Power Capacity in U.S. Increased 27% in 2006 and Is Expected to Grow an Additional 26% in 2007* [Cited April 3, 2007]. awea.org/newsroom/releases/Wind_Power_Capacity_012307.html
167. Gipe, Paul. *Rooftop Turbines: Rooftop Mounting and Building Integration of Wind Turbines* [Cited January 21, 2007]. wind-works.org/articles/RoofTopMounting.html
168. *LEED for New Construction Version 2.2 Reference Guide.* U.S. Green Building Council, p. 123. See also, Water Resources Research Center cals.arizona.edu/AZWATER/publications/sustainability/report_html/ chap5_05.html [Cited April 3, 2007].
169. Powell, Richard. *Wabi Sabi Simple.* Adams Media, 2004 [Cited January 13, 2007].

170. *Design Projects* [Cited January 13, 2007]. ecodesign.org/edi/projects/design/greengulch.html
171. BASF Press Release. January 24, 2007.
172. *Top Global Companies Join with WBCSD to Make Energy Self-Sufficient Buildings a Reality* [Cited January 13, 2007]. wbcsd.org/plugins/DocSearch/details.asp?type=DocDet&ObjectId=MTg2MTU
173. *BioRegional's Response to U.K. Government's Announcements on Zero Carbon Homes* [Cited January 13, 2007]. December 20, 2006. bioregional.com/news%20page/news_stories/ZED/zerocarbon%20201206.htm

Index

Page numbers with fig. indicate figures.

About the Author

JERRY YUDELSON has been involved in commercializing renewable energy systems, environmental remediation products and services, and green building design and consulting services for 25 years. A registered professional engineer, he holds degrees in civil and environmental engineering from Caltech and Harvard, and an MBA with highest honors from the University of Oregon.

Currently, as principal for Yudelson Associates, a green building consultancy based in Tucson, Arizona, he works to "build the business of green building." His clients include architects, developers, product manufacturers and private equity.

He has trained more than 3,000 industry professionals in the LEED system. Since 2004 he has chaired the USGBC's annual conference, Greenbuild, the largest green building conference in the world. In 2006 he was named a National Peer Professional by the US General Services Administration.

He lives with his wife and dog at the edge of the Sonoran Desert in Tucson, Arizona.

New Society Publishers

ENVIRONMENTAL BENEFITS STATEMENT

New Society Publishers has chosen to produce this book on recycled paper made with **100% post consumer waste**, processed chlorine free, and old growth free.

For every 5,000 books printed, New Society saves the following resources:[1]

24	Trees
2,188	Pounds of Solid Waste
2,408	Gallons of Water
3,140	Kilowatt Hours of Electricity
3,978	Pounds of Greenhouse Gases
17	Pounds of HAPs, VOCs, and AOX Combined
6	Cubic Yards of Landfill Space

[1]Environmental benefits are calculated based on research done by the Environmental Defense Fund and other members of the Paper Task Force who study the environmental impacts of the paper industry.

NEW SOCIETY PUBLISHERS